SOTHIS PRESS

Sothis Press aims to completely reimagine the public domain book with an emphasis on universal access to captivatingly designed print editions. We believe that technology will enable every single book ever published to be perpetually accessible in beautiful print form. Sothis is beauty in ubiquity.

www.sothis.press
1 Waterhouse Square, London, ECIN 2ST

Scientific Dialogues, Intended for the Instruction and Entertainment of Young People

In Which the First Principles of Natural and Experimental Philosophy Are Fully Explained; Of Mechanics

Jeremiah Joyce

SOTHIS PRESS

ISBN 978-1-0365-1643-7

Scientific Dialogues, Intended for the Instruction and Entertainment of Young People is in the public domain.

SCIENTIFIC DIALOGUES,

INTENDED FOR THE

INSTRUCTION AND ENTERTAINMENT

OF

YOUNG PEOPLE:

IN WHICH

THE FIRST PRINCIPLES

OF

NATURAL AND EXPERIMENTAL

PHILOSOPHY

ARE FULLY EXPLAINED.

VOL. I. OF MECHANICS.

" *Conversation, with the habit of explaining the meaning of words,*
" *and the structure of common domestic implements to children, is the*
" *sure and effectual method of preparing the mind for the acquirement of*
" *science.*" EDGEWORTH'S PRACTICAL EDUCATION.

BY THE REV. J. JOYCE.

A NEW EDITION, CORRECTED AND IMPROVED.

LONDON:

PRINTED FOR BALDWIN, CRADOCK, AND JOY,
PATERNOSTER ROW; AND
R. HUNTER, SUCCESSOR TO MR. JOHNSON,
NO. 72, ST. PAUL'S CHURCHYARD.

1818.

TO THE HONOURABLE

CHARLES BANKS STANHOPE*,

AND TO THE HONOURABLE

JAMES HAMILTON STANHOPE.

GENTLEMEN,

I AM desirous of prefixing your names to these volumes, in token of the affectionate attachment to which, from me, you are peculiarly entitled. And I am happy in the opportunity which this publi-

* This young man, the second son of the present Earl Stanhope, chose at an early period the profession of arms. In the year 1807 he was appointed aid-de-camp to General Moore, whom he accompanied to Sicily, and afterwards to Sweden. In the latter end of the year 1808, he was, as major, appointed to the command of the 50th regiment, then in Spain, at the head of which he was shot through the heart, at Corunna, on the 16th of January, 1809.

cation affords me of bringing to your recollection subjects, in the study of which you successfully engaged at a very early period of life, and which are of acknowledged importance in the pursuits of every well educated youth.

In perusing this little work you must bear in your minds, that it is not intended for proficients in philosophical knowledge, but for noviciates in science:—not for yourselves in the present advanced stage of your progress, but for those young persons who are unacquainted with the rudiments of natural and experimental philosophy.

I am too well acquainted with the excellence of your dispositions to suppose it necessary for me to apologize for laying before you a work that has no extraordinary claim to your acceptance. You will, I am sure, appreciate its value, not so much by its intrinsic contents, as by the good-will with which it is presented.

Before I conclude this short address, permit me to say, that my own happiness will ever be much augmented, by the assurance of the happiness and distinguished usefulness of those with whom I have spent so many years of my life, and to whose permanent interest, I am sure, you will acknowledge I have never been inattentive.

Sincerely wishing you, Gentlemen, all the felicity which the honourable exercise of distinguished talents and virtuous minds can confer upon the possessors,

I subscribe myself,

Your very affectionate Friend,

and obedient Servant,

THE AUTHOR.

Clapton, May, 1800.

PREFACE.

THE Author of these little volumes feels himself extremely happy in the opportunity which this publication affords him of acknowledging the obligations he is under to the authors of "Practical Education," for the pleasure and instruction which he has derived from that valuable work. To this he is indebted for the idea of writing on the subject of Natural Philosophy for the use of children. How far his plan corresponds with that suggested by Mr. Edgeworth, in his chapter on Mechanics, must be left with a candid public to decide.

The Author conceives, at least, he shall be justified in asserting, that no introduction to natural and experimental philosophy has been attempted in a method so familiar and easy as that which he now offers to the public—none which appears to him so properly adapted to the capacities of young people of ten or eleven years of age; a period of life, which, from the Author's own experience, he is confident is by no means too early to induce in children habits of scientific reasoning. In this opinion he is sanctioned by the authority of Mr. Edgeworth. "Parents," says he, "are anxious that children should be conversant with mechanics, and with what are called the mechanical powers. Certainly no species of knowledge is better suited to the taste and capacity of youth, and yet it seldom forms a part of early instruction. Every body talks of the lever, the wedge, and the pulley, but most people perceive, that the notions which they have of their respective uses are unsatisfactory and indistinct; and many endeavour, at a late period of life, to acquire a scientific and exact knowledge of the effects that are produced by implements that are in every body's hands, or that are absolutely necessary in the daily occupations of mankind."

The Author trusts, that the whole work will be found a complete compendium of natural and experimental philosophy, not only adapted to the understandings of young people, but well calculated also to convey that kind of familiar instruction which is absolutely necessary, before a person can attend public lectures in these branches of science with advantage. "If," says Mr. Edgeworth, speaking on this subject, "the lecturer does not communicate much of that knowledge which he endeavours to explain, it is not to be attributed either to his want of skill or to the insufficiency of his apparatus, but to the novelty of the terms which he is obliged to use. Ignorance of the language in which any science is taught is an insuperable bar to its being suddenly acquired; besides a precise knowledge of the meaning of terms, we must have an instantaneous idea excited in our minds whenever they are repeated; and, as this can be acquired only by practice, it is impossible that philosophical lectures can be of much service to those who are not familiarly acquainted with the technical language in which they are delivered *.

It is presumed, that an attentive perusal of these Dialogues, in which the principal and most common terms of science are carefully explained, and illustrated by a variety of familiar examples, will be the means of obviating this objection, with respect to persons who may be desirous of attending those public philosophical lectures to which the inhabitants of the metropolis have almost constant access.

* Mr. Edgeworth's chapter on Mechanics should be recommended to the attention of the reader, but the Author feels unwilling to refer to a part of a work, the whole of which deserves the careful perusal of all persons engaged in the education of youth.

CONTENTS

TO

VOL. I.

Just Published,

BY THE SAME AUTHOR.

1. The COMPANION TO THE SCIENTIFIC DIALOGUES, or PUPIL'S MANUAL IN NATURAL AND EXPERIMENTAL PHILOSOPHY; containing a complete Set of Questions, and other Exercises, for the Examination of Pupils in the Six Volumes of the Scientific Dialogues. Second Edition. Price 2s. 6d.

2. DIALOGUES IN CHEMISTRY, upon the same Plan with the SCIENTIFIC DIALOGUES; in which the first Principles of that Science are fully explained. To which are added Questions and other Exercises, for the Examination of Pupils. In Two Volumes. Third Edition. Price 7s. half bound.

3. DIALOGUES ON THE MICROSCOPE, in Two Volumes. Price 7s.

4. LETTERS ON NATURAL AND EXPERIMENTAL PHILOSOPHY, Chemistry, Anatomy, Physiology, and other Branches of Science, pertaining to the Material World. Addressed to a Youth settling in the Metropolis. Illustrated with Nineteen Plates. Price 10s. 6d.

5. DR. ASHWORTH'S INTRODUCTION TO PLANE TRIGONOMETRY; with the Application of it to the measuring of Heights and Distances, and to Land Surveying; and the Use of the Portable Case of Mathematical Instruments. A new Edition. Price 3s. 6d.

To this Edition are added, various Examples, adapted to the Use of Schools, and calculated for Persons who may wish to survey and measure Estates.

CONVERSATION I.

INTRODUCTION.

FATHER—CHARLES—EMMA

CHARLES. Father, you told sister Emma and me, that, after we had finished reading the "*Evenings at Home*," you would explain to us some of the principles of natural philosophy: will you begin this morning?

Father. Yes, I am quite at leisure; and I shall indeed at all times take a delight in communicating to you the elements of useful knowledge;

and the more so in proportion to the desire which you have of collecting and storing those facts that may enable you to understand the operations of nature, as well as the works of ingenious artists. These, I trust, will lead you, insensibly, to admire the wisdom and goodness by means of which the whole system of the universe is constructed and supported.

Emma. But can philosophy be comprehended by children so young as we are? I thought that it had been the business of men, and of old men too.

Father. Philosophy is a word which in its original sense signifies only a love or desire of wisdom; and you will not allow that you and your brother are too young to wish for knowledge.

Emma. So far from it, that the more knowledge I get, the better I seem to like it; and the number of new ideas which, with a little of your assistance, I have obtained from the "*Evenings at Home,*" and the great pleasure which I have received from the perusal of these volumes, will, I am sure, excite me to read them again and again.

Father. You will find very little in the introductory parts of natural and experimental philosophy, that requires much more of your attention than many parts of that work with which you have been so delighted. Besides, the study of natural philosophy improves and elevates the mind, by unfolding the magnificence, the order, and the beauty manifested in the construction of the material world; while it offers the most striking proofs of the beneficence, the wisdom, and the power of the Creator.

Charles. But in some books of natural philosophy, which I have occasionally looked into, a number of new and uncommon words have perplexed me; I have also seen references to figures by means of large letters and small, the use of which I did not comprehend.

Father. It is frequently a dangerous practice for young minds to dip into subjects before they are prepared, by some previous knowledge, to enter upon them; since it may create a distaste for the most interesting topics. Thus those books which you now read with so much pleasure would not have afforded you the smallest entertainment a few years ago, when you must have spelt out almost every word in each page.

The same sort of disgust will naturally be felt by persons who attempt to read works of science before the leading terms are explained and understood. The word *angle* is continually recurring in subjects of this sort; do you know what an angle is?

Emma. I do not think I do; will you explain what it means?

Father. An *angle* is made by the opening of two straight* lines. In this figure (Plate 1. Fig. 1.) there are two straight lines A B and C B meeting at the point B, and the opening made by them is called an angle.

Charles. Whether that opening be small or great, is it still called an angle?

Father. It is; your drawing com-

* *Straight* lines, in works of science, are usually denominated *right* lines.

passes may familiarise to your mind the idea of an angle; the lines in this figure will aptly represent the legs of the compasses, and the point B the joint upon which they move or turn. Now you may open the legs to any distance you please, even so far that they shall form one straight line; in that position only they do *not* form an angle. In every other situation an angle is made by the opening of these legs, and the angle is said to be greater or less, as that opening is greater or less.

Emma. Are not some angles called right angles?

Father. Angles are either *right*, *acute*, or *obtuse*. When the line A B (Plate 1. Fig. 2.) meets another line B C, in such a manner as to make the angles A B D and A B C equal to one another, then those angles are called *right* angles. And the line A B is said to be perpendicular to D C. Hence to be perpendicular to, or to make *right* angles with a line, means one and the same thing.

Charles. Does it signify how you call the letters of an angle?

Father. It is usual to call every angle by three letters, and that at the angular point must be always the middle letter of the three. There are cases, however, where an angle may be denominated by a single letter, as in figures 1 and 3, the angle A B C may be called simply the angle B, for in these figures there is no danger of mistake, because there is but a single angle at the point B.

Charles. I understand this, for if in the second figure I were to describe the angle by the letter B only, you would not know whether I meant the angle A B C or A B D.

Father. That is the precise reason why it is necessary in most descriptions to make use of three letters. An *acute* angle (Fig. 1.) A B C is less than a right angle; and an *obtuse* angle (Fig. 3.) A B C is greater than a right angle.

Emma. You see the reason now, Charles, why letters are placed against or by the figures, which puzzled you before.

Charles. I do; they are intended to distinguish the separate parts of each, in order to render the description of them easier both to the author and the reader.

Emma. What is the difference between an angle and a triangle?

Father. An angle being made

by the opening of two lines, and, as you know, that two straight lines cannot enclose a space, so a *triangle* ABC (Fig. 4.) is a space bounded by three straight lines. It takes its name from the property of containing three angles. There are various sorts of triangles; but it is not necessary to enter upon these particulars, as I do not wish to burthen your memories with more technical terms than we have occasion for.

Charles. A triangle then is a space or figure containing three angles, and bounded by as many straight lines.

Father. Yes, that description will answer our present purpose.

CONVERSATION II.

Of Matter.—Of the Divisibility of Matter.

FATHER. Do you understand what philosophers mean when they make use of the word matter?

Emma. Are not all things which we see and feel composed of matter?

Father. Every thing which is the object of our senses is composed of matter differently modified or arranged. But in a philosophical sense *matter* is defined to be an *extended, solid, inactive,* and *moveable* substance.

Charles. If by extension is meant length, breadth, and thickness, matter, undoubtedly, is an extended substance. Its solidity is also manifest by the resistance it makes to the touch.

Emma. And the other properties nobody will deny, for all material objects are, of themselves, without motion; and yet it may be readily conceived, that by the application of a proper force there is no body which cannot be moved. But I remember, that you told us something strange about the divisibility of matter, which you said might be continued without end.

Father. I did, some time ago, mention this as a curious and interesting subject, and this is a very fit time for me to explain it.

Charles. Can matter indeed be

infinitely divided, for I suppose that this is what is meant by a division without end?

Father. Difficult as this may at first appear, yet I think it very capable of proof. Can you conceive of a particle of matter so small as not to have an upper and under surface?

Charles. Certainly, every portion of matter, however minute, must have two surfaces at least, and then I see, that it follows of course that it is divisible; that is, the upper surface may be separated from the lower.

Father. Your conclusion is just; and though there may be particles of matter too small for us actually to divide, yet this arises from the imperfection of our instruments; they must nevertheless, in their nature, be divisible.

Emma. But you were to give us some remarkable instances of the minute division of matter.

Father. A few years ago a lady spun a single pound of wool into a thread 168 thousand yards long. And Mr. Boyle mentions, that two grains and a half of silk were spun into a thread 300 yards in length. If a pound of silver, which, you know, contains 5760 grains, and a single grain of gold be melted together, the gold will be equally diffused through the whole silver, insomuch that if one grain of the mass be dissolved in a liquid called *Aqua Fortis*, which is diluted nitric acid, the gold will fall to the bottom. By this experiment it is evident that a grain may be divided into 5761 visible parts, for only the 5761st part of the gold is contained in a single grain of the mass.

The gold-beaters, whom you have seen at work in the shops in Long Acre, can spread a grain of gold into a leaf containing 50 square inches, and this leaf may be readily divided into 500,000 parts, each of which is visible to the naked eye: and by the help of a microscope, which magnifies the area or surface of a body 100 times, 100th part of each of these becomes visible; that is, the 50 millionth part of a grain of gold will be visible, or a single grain of that metal may be divided into 50 million of visible parts. But the gold which covers the silver wire used in making what is called gold lace, is spread over a much larger surface, yet it preserves, even if examined by a microscope, an uniform appearance. It has been calculated that one grain of gold under these circumstances would cover a surface of nearly thirty square yards.

The *natural* divisions of matter are still more surprising. In odoriferous bodies, such as camphor, musk, and asafœtida, a wonderful subtilty of parts is perceived, for though they are perpetually filling a considerable space with odoriferous particles, yet these bodies lose but a very small part of their weight in a great length of time.

Again, it is said by those who have examined the subject with the best glasses, and whose accuracy may be relied on, that there are more animals in the milt of a single cod-fish, than there are men on the whole earth, and that a single grain of sand is larger than four millions of these animals. Now if it be admitted that these little animals are possessed of organized parts, such as a heart, stomach, muscles, veins, arteries, &c., and that they are possessed of a complete system of circulating fluids, similar to what is found in larger animals, we seem to approach to an idea of the infinite divisibility of matter. It has indeed been calculated that a particle of blood of one of these animalcula is as much smaller than a globe one-tenth of an inch in diameter, as that globe is smaller than the whole earth. Nevertheless, if these particles be compared with the particles of light, it is probable, that they would be found to exceed them in bulk as much as mountains do single grains of sand:—

In thousand species of the insect kind,
Lost to the naked eye, so wondrous small,
Were millions join'd, one grain of sand would
cover all.
Yet each, within its little bulk, contains
A heart which drives the torrent through its
veins;
Muscles to move its limbs aright: a brain
And nerves dispos'd for pleasure and for pain;
Eyes to distinguish: sense whereby to know
What's good or bad; is, or is not its foe.

BAKER.

I might enumerate many other instances of the same kind; but these, I doubt not, will be sufficient to convince you into what very minute parts matter is capable of being divided; and with these we will put an end to our present conversation.

CONVERSATION III.

Of the Attraction of Cohesion.

FATHER. Well, my children, have you reflected upon what we last conversed about? Do you comprehend the several instances which I enumerated as examples of the minute division of matter?

Emma. Indeed, the examples which you gave us very much excited my wonder and admiration, and yet, from the thinness of some leaf-gold which I once had, I can readily admit all you have said on that part of the subject. But I know not how to conceive of such small animals as you described; and I am still more puzzled in imagining, that animals so minute actually possess all the properties of the larger ones, such as a heart, veins, blood, &c.

Father. I can, the next bright morning, by the help of the solar microscope, show you very distinctly, the circulation of the blood in a flea, which you may get from your little dog; and with better glasses than those of which I am possessed, the same might be shown in animals still smaller than the flea, perhaps, even in those which are themselves invisible to the naked eye. But we shall converse more at large on this matter when we come to consider the subject of optics, and the construction and uses of the solar microscope. At present we will turn our thoughts to that principle in nature, which philosophers have agreed to call gravity or attraction.

Charles. If there be no more difficulties in philosophy than we met with in our last lecture, I do not fear but that we shall, in general, be able to understand it. Are there not several kinds of gravity?

Father. Yes, there are; two of which it will be sufficient for our present purpose to describe; the one is the *attraction of cohesion;* the other that of *gravitation.* The *attraction of cohesion* is that power which keeps the parts of bodies together when they touch, and prevents them from separating, or which inclines the parts of bodies to unite, when they are placed sufficiently near to each other.

Charles. Is it then by the attraction of cohesion that the parts of this table, or of the pen-knife, are kept together?

Father. The instances which you have selected are accurate, but you might have said the same of every other solid substance in the room; and it is in proportion to the different degrees of attraction with which different substances are affected, that some bodies are hard, others soft, tough, &c. A philosopher in Holland, almost a century ago, took great pains in ascertaining the different degrees of cohesion, which belonged to various kinds of wood, metals, and many other substances. A short account of the experiments made by M. Musschenbroek, you will hereafter find in your own language, in the second edition of Dr. Enfield's Institutes of Natural Philosophy.

Charles. You once showed me that two leaden bullets, having a little scraped from the surfaces, might be made, with a sort of twisting pressure, to stick together with great force; you called that, I believe, the attraction of cohesion?

Father. I did: some philosophers, who have made this experiment with great attention and accuracy, assert, that if the flat surfaces, which are presented to one another, be but a quarter of an inch in diameter, scraped very smooth, and forcibly pressed together with a twist, a weight of a hundred pounds is frequently required to separate them.

As it is by this kind of attraction

that the parts of solid bodies are kept together, so when any substance is separated or broken, it is only the attraction of cohesion that is overcome in that particular part.

Emma. Then when I had the misfortune this morning at breakfast, to let my saucer slip from my hands, by which it was broken into several pieces, was it only the attraction of cohesion that was overcome by the parts of the saucer being separated in its fall on the ground?

Father. Just so; for whether you unluckily break the china, or cut a stick with your knife, or melt lead over the fire, as your brother sometimes does, in order to make plummets; these, and a thousand other instances, which are continually occurring, are but examples in which the cohesion is overcome by the fall, the knife, or the fire.

Emma. The broken saucer being highly valued by mamma, she has taken the pains to join it again with white lead; was this performed by means of the attraction of cohesion?

Father. It was, my dear; and hence you will easily learn that many operations in cookery are in fact nothing more than different methods of causing this attraction to take place. Thus flour, by itself, has little or nothing of this principle; but when mixed with milk, or other liquids, to a proper consistency, the parts cohere strongly; and this cohesion in many instances becomes still stronger, by means of the heat applied to it in boiling or baking.

Charles. You put me in mind of the fable of the man blowing hot and cold; for in the instance of the *lead*, fire overcomes the attraction of cohesion; and the same power, heat, when applied to puddings, bread, &c., causes their parts to cohere more powerfully. How are we to understand this?

Father. I will endeavour to remove your difficulty. Heat expands all bodies without exception, as you shall see before we have finished our lectures. Now the fire applied to metals in order to melt them, causes such an expansion, that the particles are thrown out of the sphere, or reach of each other's attraction: whereas the heat communicated in the operations of cookery, is sufficient to expand the particles of flour, but is not enough to overcome the attraction of cohesion. Besides, your mother will tell you, that the heat of boiling would frequently disunite the parts of which her puddings are composed, if she did not take the precaution of enclosing them in a cloth, leaving them just room enough to expand without the liberty of breaking to pieces; and the moment they are taken from the water, they lose their superabundant heat, and become solid.

Emma. When the cook makes broth for little brother, it is the heat then which overcomes the attraction which the particles of meat have for each other, for I have seen her pour off the broth, and the meat is all in rags. But will not the heat overcome the attraction which the parts of the bones have for each other?

Father. The heat of boiling water will never effect this; but a machine was invented several years ago, by Mr. Papin, for that purpose. It is called Papin's Digester, and is used in taverns, and in many large families, for the purpose of dissolving bones, as completely as a lesser degree of heat will liquefy jelly. On some future day I will show you an engraving of this machine, and explain its different parts, which are extremely simple*.

* See Vol. IV. Conver. XIX. p. 185—8.

CONVERSATION IV.

Of the Attraction of Cohesion.

FATHER. I will now mention some other instances of this great law of nature. If two polished plates of marble, or brass, be put together, with a little oil between them to fill up the pores of their surfaces, they will cohere so powerfully as to require a very considerable force to separate them.—Two globules of quicksilver, placed very near to each other, will run together and form one large drop.—Drops of water will do the same.—Two circular pieces of cork placed upon water at about an inch distant will run together.—Balance a piece of smooth board on the end of a scale beam; then let it lie flat on water, and five or six times its own weight will be required to separate it from the water. If a small globule of quicksilver be laid on clean paper, and a piece of glass be brought into contact with it, the mercury will adhere to it, and be drawn away from the paper. But bring a larger globule into contact with the smaller one, and it will forsake the glass, and unite with the other quicksilver.

Charles. Is it not by means of the attraction of cohesion, that the little tea which is generally left at the bottom of the cup instantly ascends in the sugar when thrown into it?

Father. The ascent of water or other liquids in sugar, sponge, and all porous bodies, is a species of this attraction, and is called *capillary** *attraction;* it is thus denominated from the property which tubes of a very small bore, scarcely larger than to admit a hair, have of causing water to stand above its level.

Charles. Is this property visible in no other tubes than those, the bores of which are so exceedingly fine?

Father. Yes, it is very apparent in tubes whose diameters are one tenth of an inch or more in length; but the smaller the bore, the higher the fluid rises; for it ascends, in all instances, till the weight of the co-

* From *capillus,* the Latin word for *hair.*

lumn of water in the tube balances, or is equal to the attraction of the tube. By immersing tubes of different bores in a vessel of coloured water, you will see that the water rises as much higher in the smaller tube, than in the larger, as its bore is less than that of the larger. The water will rise a quarter of an inch, and there remain suspended in a tube, whose bore is about one eighth of an inch in diameter.

This kind of attraction is well illustrated by taking (Plate 1. Fig. 5.) two pieces of glass joined together at the side B C, and kept a little open at the opposite side A D, by a small piece of cork E. In this position immerse them in a dish of coloured water F G, and you will observe that the attraction of the glass at, and near B C, will cause the fluid to ascend to B, whereas about the parts D it scarcely rises above the level of the water in the vessel.

Charles. I see that a curve is formed by the water.

Father. There is, and to this curve there are many curious properties belonging, as you will hereafter be able to investigate for yourself.

Emma. Is it not upon the principle of the attraction of cohesion, that carpenters glue their work together?

Father. It is upon this principle that carpenters and cabinet-makers make use of glue; that braziers, tinmen, plumbers, &c., solder their metals; and that smiths unite different bars of iron by means of heat. These and a thousand other operations, of which we are continually the witnesses, depend on the same principle as that which induced your mother to use the white lead in mending her saucer. And you ought to be told, that though white lead is frequently used as a cement for broken china, glass, and earthen ware, yet if the vessels are to be brought again into use, it is not a proper cement, being an active poison; besides, one much stronger has been discovered, I believe, by a very able and ingenious philosopher, the late Dr. Ingenhouz, at least I had it from him several years ago; it consists simply of a mixture of quick-lime, and Glocester cheese, rendered soft by warm water, and worked up to a proper consistency.

Emma. What! do such great philosophers, as I have heard you say

Dr. Ingenhouz was, attend to such trifling things as these?

Father. He was a man deeply skilled in many branches of science; and I hope that you and your brother will one day make yourselves acquainted with many of his important discoveries. But no real philosopher will consider it beneath his attention to add to the conveniencies of life.

Charles. This attraction of cohesion seems to pervade the whole of nature.

Father. It does, but you will not forget that it acts only at very small distances. Some bodies indeed appear to possess a power the reverse of the attraction of cohesion.

Emma. What is that?

Father. It is called repulsion.—

Thus water repels most bodies till they are wet. A small needle carefully placed on water will swim, although the iron of which it is made is much heavier than water: flies walk upon water without wetting their feet:—

> Or bathe unwet their oily forms, and dwell
> With feet repulsive on the dimpling well.
>
> DARWIN.

The drops of dew which appear in a morning on plants, particularly on cabbage plants, assume a globular form, from the mutual attraction between the particles of water; and upon examination it will be found that the drops do not touch the leaves, for they will roll off in compact bodies, which could not be the case if there subsisted any degree of attraction between the water and the leaf.

If a small thin piece of iron be laid upon quicksilver, the repulsion between the different metals will cause the surface of the quicksilver near the iron to be depressed.

The repelling force of the particles of a fluid is but small; therefore if a fluid be divided it easily unites again. But if a glass or any hard substance be broken, the parts cannot be made to cohere without first being moistened, because the repulsion is too great to admit of a reunion.

The repelling force between water and oil is likewise so great, that it is impossible to mix them in such a manner that they shall not separate again.

If a ball of light wood be dipped in oil, and then put into water, the water will recede so as to form a sort of channel around the ball.

Charles. Why do cane, steel, and many other things, bear to be bent without breaking, and, when set at liberty, recover their original form.

Father. That a piece of thin steel, or cane, recovers its usual form, after being bent, is owing to a certain power called *elasticity;* which may, perhaps, arise from the particles of those bodies, though disturbed, not being drawn out of each other's attraction; therefore, as soon as the force upon them ceases to act, they restore themselves to their former position.—But our half hour is expired, I must leave you.

CONVERSATION V

Of the Attraction of Gravitation.

FATHER. We will now proceed to discuss another very important general principle in nature; the *attraction of gravitation*, or, as it is frequently termed, *gravity*, which is that power by which *distant* bodies tend towards each other. Of this we have perpetual instances in the falling of bodies to the earth.

Charles. Am I then to understand, that whether this marble falls from my hand; or a loose brick from the top of the house; or an apple from the tree in the orchard, that all these happen by the attraction of gravity?

Father. It is by the power which is commonly expressed under the term *gravity*, that all bodies whatever have a tendency to the earth, and, unless supported, will fall in lines nearly perpendicular to its surface. This tendency or disposition to fall is called weight: and the weight of a given body may be employed to measure the weights of all other bodies. Gravity is to be distinguished from weight. The weight of a body is the product of the gravity into the number of particles of which the body is composed.

Emma. But are not smoke, steam, and other light bodies, which we see ascend, exceptions to the general rule?

Father. It appears so at first sight; and it was formerly received as a general opinion, that smoke, steam, &c., possessed no weight: the discovery of the air-pump has shown the fallacy of this notion, for in an exhausted receiver, that is, in a glass jar from which the air is taken away by means of the air-pump, smoke and steam descend by their own weight as completely as a piece of lead. When we come to converse on the subject of pneumatics and hydrostatics, you will understand that the reason why smoke and other bodies ascend is simply because they are lighter than the atmosphere which surrounds them, and the moment they reach that part of it which has the same gravity with themselves they cease to rise.

Charles. Is it then by this power

that all terrestrial bodies remain firm on the earth?

Father. By gravity, bodies on all parts of the earth (which you know is of a globular form) are kept on its surface, because they all, wherever situated, tend to the centre; and, since all have a tendency to the centre, the inhabitants of New Zealand, although nearly opposite to our feet, stand as firm as we do in Great Britain.

Charles. This is difficult to comprehend: nevertheless, if bodies on all parts of the surface of the earth have a tendency to the centre, there seems no reason why bodies should not stand firm on one part as well as another. Does this power of gravity act alike on all bodies?

Father. It does, without any regard to their figure, or size; for attraction or gravity acts upon bodies in proportion to the quantity of matter which they contain, that is, four times a greater force of gravity is exerted upon a weight of four pounds, than upon one of a single pound. The consequence of this principle is, that all bodies at equal distances from the earth fall with equal velocity.

Emma. What do you mean, papa, by *velocity?*

Father. I will explain it by an example or two; if you and Charles set out together, and *you* walk a mile in half an hour, but *he* walk and run two miles in the same time, how much swifter will he go than you?

Emma. Twice as swift.

Father. He does, because, *in the*

same time, he passes over twice as much space; therefore we say his velocity is twice as great as yours. Suppose a ball, fired from a cannon, pass through 800 feet in a second of time; and, in the same time, your brother's arrow pass through 100 feet only, how much swifter does the cannon ball fly than the arrow?

Emma. Eight times swifter.

Father. Then it has eight times the velocity of the arrow; and hence you understand that swiftness and velocity are synonymous terms, and that the velocity of a body is measured by the space it passes over in a given time, as a second, a minute, an hour, &c.

Emma. If I let a piece of metal, as a penny-piece, and a feather fall from my hand at the same time, the penny will reach the ground much sooner than the feather. Now how do you account for this, if all bodies are equally affected by gravitation, and descend with equal velocities, when at the same distance from the earth?

Father. Though the penny and feather will not, in the open air, fall with equal velocity, yet, if the air be taken away, which is easily done, by a little apparatus connected with the air-pump, they will descend in the same time. Therefore the true reason why light and heavy bodies do not fall with equal velocities is, that the *former*, in proportion to its weight, meets with a much greater resistance from the air than the *latter*.

Charles. It is then, I imagine,

from the same cause, that, if I drop the penny and a piece of light wood into a vessel of water, the penny shall reach the bottom, but the wood, after descending a small way, rises to the surface.

Father. In this case the resisting medium is water instead of air, and the copper, being about nine times heavier than its bulk of water, falls to the bottom without apparent resistance. But the wood, being much lighter than water, cannot sink in it; therefore, though by its *momentum** it sinks a small distance, yet as soon as that is overcome by the resisting medium, that is the water, it rises to the surface, being the lighter substance.

* The explanation of this term will be found in the next Conversation.

CONVERSATION VI.

Of the Attraction of Gravitation.

EMMA. The term *momentum*, which you made use of yesterday, is another word which I do not understand.

Father. If you have understood what I have said respecting the velocity of moving bodies, you will easily comprehend what is meant by the word momentum.

The *momentum*, or moving force of a body, is measured by its weight multiplied into its velocity. You may, for instance, place this pound weight upon a china plate without any danger of breaking, but if you let it fall from the height of only a few inches, it will dash the china to pieces. In the first case, the plate has only the pound weight to sustain, in the other the weight must be multiplied into the velocity, or, to speak in a popular manner, into the distance of the height from which it fell.

If a ball *a* (Plate 1. Fig. 6.) lean against the obstacle *b*, it will not be able to overturn it, but if it be taken up to *c* and suffered to roll down the inclined plane A B against *b* it will certainly overthrow it;—in the former case *b* would only have to resist the weight of the ball *a*, in the latter it has to resist the weight multiplied into its motion or velocity.

Charles. Then the momentum of

a small body, whose velocity is very great, may be equal to that of a very large body with a slow velocity.

Father. It may; and hence you see the reason why immense battering rams, used by the ancients, in the art of war, have given place to cannon balls of but a few pounds weight.

Charles. I do; for what is wanting in weight is made up by velocity.

Father. Can you tell me what velocity a cannon ball of 28 pounds must have to effect the same purposes as would be produced by a battering ram of 15,000 pounds weight, and which, by manual strength, could be moved at the rate of only two feet in a second of time?

Charles. I think I can;—the *mo-*

mentum of the battering ram must be estimated by its weight, multiplied into the space passed over in a second, which is 15,000 multiplied by two feet equal to 30,000; now if this momentum, which must also be that of the cannon ball, be divided by the weight of the ball, it will give the velocity required; and 30,000 divided by 28 will give for the quotient 1072 nearly, which is the number of feet which the cannon ball must pass over in a second of time in order that the momenta of the battering ram and the ball may be equal, or in other words, that they may have the same effect in beating down an enemy's wall.

Emma. I now fully comprehend what the momentnm of a body is, 'or if I let a common trap-ball accidentally fall from my hand, upon my foot, it occasions more pain than the mere pressure of a weight several times heavier than the ball.

Charles. If the attraction of gravitation be a power by which bodies in general tend towards each other, why do all bodies tend to the earth as a centre?

Father. I have already told you, that, by the great law of gravitation, the attraction of all bodies is in proportion to the quantity of matter which they contain. Now the earth, being so immensely large in comparison of all other substances in its vicinity, destroys the effect of this attraction between smaller bodies, by bringing them all to itself—If two balls are let fall from a high tower at a small distance apart; though they have an attraction for one another, yet it will be as nothing when compared with the attraction by which they are both impelled to the earth, and consequently the tendency which they mutually have of approaching one another will not be perceived in the fall. If, however, any two bodies were placed in free space, and out of the sphere of the earth's attraction, they would, in that case, assuredly approach each other, and that with increased velocity as they came nearer. If the bodies were equal, they would meet in the middle point between the two; but if they were unequal, they would then meet as much nearer the larger one, as that contained a greater quantity of matter than the other.

Charles. According to this, the

earth ought to move towards falling bodies, as well as they move to it.

Father. It ought, and, in just theory, it does; but when you calculate how many million of times larger the earth is than any thing belonging to it, and if you reckon, at the same time, the small distances from which bodies can fall, you will know that the point where the falling bodies and earth will meet, is removed only to an indefinitely small distance from its surface, a distance much too small to be conceived by the human imagination.

As all bodies on or near the earth tend to the centre of that body; so the earth, and all the planets, with their several moons, as we shall see by and by, tend to the centre of the sun, as the point to which the whole and every part of the solar system is attracted.

We will resume the subject of gravity to-morrow.

CONVERSATION VII.

Of the Attraction of Gravitation.

EMMA. Has the attraction of gravitation the same effect on all bodies, whatever be their distance from the earth?

Father. No; this, like every power which proceeds from a centre, decreases as the squares of the distances from that centre increase.

Emma. I fear that I shall not understand this, unless you illustrate it by examples.

Father. Suppose you are reading at the distance of one foot from a candle, and that you receive a certain quantity of light on your book; now if you remove to the distance of two feet from the candle, you will, by this law, receive four times less light than you had before; here then, though you have increased your distance but two-fold, yet the light is diminished four-fold, because four is the square of two, or two multiplied by itself. If, instead of removing two feet from the candle, you take your station at 3, 4, 5, or 6 feet distance, you will then receive at the different distances, 9, 16, 25, 36 times less light than when you were within a single foot from the candle, for these, as you know, are the squares of the numbers 3, 4, 5, and 6. The same is applicable to the heat imparted by a fire; at the distance of one yard from which, a person will enjoy four times as much heat as he who sits or stands two yards from it; and nine times as much as one that shall be removed to the distance of three yards.

Charles. Is then the attraction of gravity four times less at a yard distance from the earth than it is at the surface?

Father. No; whatever be the cause of attraction, which to this day remains undiscovered, it acts from the *centre* of the earth, and not from its surface, and hence the difference of the power of gravity cannot be discerned at the small distances to which we can have access; for a mile or two, which is much higher than, in general, we have opportunities of making experiments, is nothing in comparison of 4000 miles, the distance of the centre from the surface of the earth. But could we ascend 4000 miles above the earth, and of course be double the distance that we now are from the centre, we should there find that the attractive force would be but one fourth of what it is here; or, in other words, that a body, which, at the surface of the earth, weighs one pound, and by the force of gravity falls through sixteen feet in a second of time, would at 4000 miles above the earth weigh but a quarter of a pound, and fall through only four feet in a second*.

* Ex. Suppose it were required to find the weight of a leaden ball, at the top of a mountain three miles high, which on the surface of the earth weighs 20lb.

If the semi-diameter of the earth be taken at 4000; then add to this the height of the

Emma. How is that known, papa, for nobody ever was there?

Father. You are right, my dear, for Garnerin, who last summer astonished all the people of the metropolis and its neighbourhood, by his flight in a balloon, ascended but a little way in comparison of the distance that we are speaking of. However, I will try to explain in what manner philosophers have come by their knowledge on this subject.

The moon is a heavy body connected with the earth by this bond of attraction, and, by the most accurate observations, it is known to be obe-

mountain, and say as the square of 4003 is to the square of 4000, so is 20lb. to a fourth proportional: or as 16024009 : 16000000 : : 20 : 19.97, or something more than 19lb. 15½oz. which is the weight of the leaden ball at the top of the mountain.

dient to the same laws as other heavy bodies are: its distance is also clearly ascertained, being about 240,000 miles, or equal to about sixty semi-diameters of the earth, and of course the earth's attraction upon the moon ought to diminish in the proportion of the square of this distance, that is, it ought to be 60 times 60, or 3600 times less at the moon than it is at the surface of the earth. This is found to be the case.

Again, the earth is not a perfect sphere, but a spheroid, that is, of the shape of an orange, rather flat at the two ends called the poles, and the distance from the centre to the poles is about eighteen or nineteen miles less than its distance from the centre to the equator; consequently, bodies ought to be something heavier at, and near the poles, than they are at the equator, which is also found to be the case. Hence it is inferred, that the attraction of gravitation varies at all distances from the centre of the earth, in proportion as the squares of those distances increase*.

Charles. It seems very surprising that philosophers who have discovered so many things have not been able to find out the cause of gravity. Had sir Isaac Newton been asked why a marble, dropped from the hand, falls to the ground, could he not have assigned a reason?

Father. That great man, probably the greatest man that ever adorned this world, was as modest as he was great, and he would have told you he knew not the cause.

* See Vol. II. Conver. VI.

The excellent and learned Dr. Price, in a work which he published thirty years ago, asks, "who does not remember a time when he would have wondered at the question, *why does water run down hill?* What ignorant man is there who is not persuaded that he understands this perfectly? But every *improved* man knows it to be a question he cannot answer." For the descent of water, like that of other heavy bodies, depends upon the attraction of gravitation, the cause of which is still involved in darkness.

Emma. You just now said that heavy bodies by the force of gravity fall sixteen feet in a second of time; is that always the case?

Father. Yes, all bodies near the surface of the earth fall at that rate in the first second of time, but as the attraction of gravitation is continually acting, so the velocity of falling bodies is an increasing, or, as it is usually called, an *accelerating* velocity. It is found by very accurate experiments that a body, descending from a considerable height by the force of gravity, falls 16 feet in the first second of time; 3 times 16 feet in the next; 5 times 16 feet in the third; 7 times sixteen in the fourth second of time; and so on, continually increasing according to the odd numbers, 1, 3, 5, 7, 9, 11, &c.

CONVERSATION VIII.

Of the Attraction of Gravitation.

EMMA. And would a ball of twenty pounds weight here, weigh half an ounce less on the top of the mountain?

Father. Certainly; but you would not be able to ascertain it by means of a pair of scales and another weight, because both weights being in similar situations would lose equal portions of their gravity.

Emma. How, then, would you make the experiment?

Father. By means of one of those

steel spiral-spring instruments which you have seen occasionally used, the fact might be ascertained.

Charles. I think, from what you told us yesterday, that with the assistance of your stop-watch, I could tell the height of any place, by observing the number of seconds that a marble or other heavy body would take in falling from that height.

Father. How would you perform the calculation?

Charles. I should go through the multiplications according to the number of seconds, and then add them together.

Father. Explain yourself more particularly; supposing you were to let a marble or penny-piece fall down that deep well which we saw last summer in the brick-field near Ramsgate, and that it was exactly five seconds in the descent, what would be the depth of the well?

Charles. In the first second it would fall 16 feet; in the next 3 times 16 or 48 feet; in the third 5 times 16 or 80 feet; in the fourth 7 times 16 or 112 feet; and in the fifth second 9 times 16 or 144 feet; now if I add 16, 48, 80, 112, and 144 together, the sum will be 400 feet, which, according to your rule, is the depth of the well. But was the well so deep?

Father. I do not think it was, but we did not make the experiment; should we ever go to that place again, you may satisfy your curiosity. You recollect that at Dover Castle we were told of a well there 360 feet deep.

Though your calculation was accurate, yet it was not done as nature effects her operations, that is in the shortest way.

Charles. I should be pleased to know an easier method; this, however, is very simple, it required nothing but multiplication and addition.

Father. True, but suppose I had given you an example in which the number of seconds had been fifty instead of five, the work would have taken you an hour or more to have performed it: whereas, by the rule which I am going to give, it might have been done in half a minute.

Charles. Pray let me have it, I hope it will be easily remembered.

Father. It will; I think it cannot be forgotten after it is once understood. The rule is this, "*the spaces described by a body falling freely from a state of rest increase as the* SQUARES *of the times increase.*" Consequently you have only to square the number of seconds, that is, to multiply the number into itself; and then multiply that again by sixteen feet, the space which it describes in the first second, and you have the required answer. Now try the example of the *well.*

Charles. The square of 5, for the time, is 25, which, multiplied by 16, gives 400, just as I brought it out before. Now if the seconds had been 50, the answer would be 50 times 50, which is 2500, and this multiplied by 16, gives 40,000 for the space required.

Father. I will now ask your sister

a question, to try how she has understood this subject. Suppose you observe by this watch that the time of the flight of your brother's arrow is exactly six seconds, to what height does it rise?

Emma. This is a different question, because here the *ascent* as well as the *fall* of the arrow is to be considered.

Father. But you will remember, that the time of the ascent is always equal to that of the descent; for as the velocity of the descent is generated by the force of gravity, so is the velocity of the ascent destroyed by the same force.

Emma. Then the arrow was three seconds only in falling; now the square of 3 is 9, which multiplied by 16, for the number of feet described in the first second, is equal to 144 feet, the height to which it rose.

Father. Now, Charles, if I get you a bow, which will carry an arrow so high as to be fourteen seconds in its flight, can you tell me the height to which it ascends?

Charles. I can now answer you without hesitation:—it will be 7 seconds in falling, the square of which is 49, and this again multiplied by 16 will give 784 feet, or rather more than 261 yards for the answer.

Father. If you will now consider the example, which you did the long way, you will see that the rule which I have given you answers very completely. In the first second the body fell 16 feet, and in the next 48, these added together make 64, which is the square of the 2 seconds multiplied by 16. The same holds true of the 3 first seconds, for in the third second it fell 80 feet, which added to the 64, give 144 equal to the square of 3 multiplied by 16. Again, in the fourth second it fell 112 feet, which added to 144, give 256 equal to the square of 4 multiplied by 16; and in the fifth second it fell 144 feet, which added to 256, give 400 equal to the square of 5 multiplied by 16. Thus you will find the rule hold in all cases, *that the spaces described by bodies falling freely from a state of rest, increase as the* **SQUARES** *of the times increase.*

Charles. I think I shall not forget the rule. I will also show my cousin Henry how he may know the height to which his bow will carry.

Father. The surest way of keeping what knowledge we have obtained, is by communicating it to our friends.

Charles. It is a very pleasant circumstance, indeed, that the giving away is the best method of keeping, for I am sure the being able to oblige one's friends is a most delightful thing.

Father. I have but a word or two more on the subject:—since the *whole spaces* described increase as the squares of the times increase, so also the *velocities* of falling bodies increase in the same proportion; for you know that the velocity must be measured by the space passed through. Thus if a person travels six miles an hour, and another person travels twelve miles in the same time, the latter will go with double the velocity of the former; consequently the *velocities* of falling bodies increase as the squares of the times increase.

If now you compare the spaces described by falling bodies in the *several moments of time taken separately*, and in their order from the beginning of the fall, then they, and consequently their velocities also, are to one another as the odd numbers, 1, 3, 5, 7, 9, 11, 13, &c., taken in their natural order, as you will observe by reflecting on the foregoing examples.

With this we conclude our present Conversation.

CONVERSATION IX.

On the Centre of Gravity.

FATHER. We are now going to treat upon the *Centre of Gravity*, which is that point of a body in which its whole weight is, as it were, concentrated, and upon which, if the body be freely suspended, it will rest; and in all other positions it will endeavour to descend to the lowest place to which it can get.

Charles. All bodies then, of whatever shape, have a centre of gravity?

Father. They have: and if you conceive a line drawn from the centre of gravity of a body towards the centre of the earth, that line is called the *line of direction,* along which every body, not supported, endeavours to fall. If the *line of direction* fall within the base of any body, it will stand; but if it does not fall within the base, the body will fall.

If I place the piece of wood A (Plate 1, Fig. 7) on the edge of a table, and from a pin *a* at its centre of gravity be hung a little weight *b*, the line of direction *ab* falls within the base, and therefore, though the wood leans, yet it stands secure. But if upon A, another piece of wood B be placed, it is evident that the centre of gravity of the whole will be now raised to *c*, at which point, if a weight be hung, it will be found that the line of direction falls out of the base, and therefore the body must fall.

Emma. I think I now see the reason of the advice which you gave me, when we were going across the Thames in a boat.

Father. I told you that if ever you were overtaken by a storm, or by a squall of wind, while you were on the water, never to let your fears so get the better of you, as to make you rise from your seat, because by so doing you would elevate the centre of gravity, and thereby, as is evident by the last experiment, increase the danger: whereas, if all the persons in the vessel were, at the moment of danger, instantly to slip from their places on to the bottom, the risque would be exceedingly diminished, by bringing the centre of gravity much lower within the vessel. The same principle is applicable to those who may be in danger of being overturned in any carriage whatever.

Emma. Surely then, papa, those stages, which load their tops with a dozen or more people, cannot be safe for the passengers.

Father. They are very unsafe; but they would be more so, were not the roads about the metropolis remarkably even and good; and, in general, it is only within twenty or thirty miles of London, or other great towns, that the tops of carriages are loaded to excess.

Charles. I understand then, that the nearer the centre of gravity is to the base of a body, the firmer it will stand.

Father. Certainly; and hence you learn the reason why conical bodies stand so sure on their bases, for the tops being small in comparison of the lower parts, the centre of gravity is thrown very low: and if the cone be upright or perpendicular, the line of direction falls in the middle of the base, which is another fundamental property of steadiness in bodies. For the broader the base, and the nearer the line of direction is to the middle of it, the more firmly does a body stand: but if the line of direction fall near the edge, the body is easily overthrown.

Charles. Is that the reason why a ball is so easily rolled along a horizontal plane?

Father. It is; for in all spherical bodies, the base is but a point, consequently almost the smallest force is sufficient to remove the line of direction out of it. Hence it is evident, that heavy bodies situated on an inclined plane will, while the line of direction falls within the base, slide down upon the plane: but they will roll when that line falls without the base. The body A (Plate I, Fig. 8) will slide down the plane DE, but the bodies B and C will roll down it.

Emma. I have seen buildings lean very much out of a straight line, why do they not fall?

Father. It does not follow because a building leans, that the centre of gravity does not fall within the base. There is a high tower at Pisa, a town in Italy, which leans fifteen feet out of the perpendicular; strangers tremble in passing by it; still it is found by experiment that the line of direction falls within the base, and therefore it will stand while its materials hold together.

A wall at Bridgenorth in Shropshire, which I have seen, stands in a similar situation, but so long as a line *cb* (Plate II, Fig. 9) let fall from the centre of gravity *c* of the building A B, passes within the base C B, it will remain firm, unless the materials with which it is built go to decay.

Charles. It must be of great use in many cases to know the method of finding the centre of gravity in different kinds of bodies.

Father. There are many easy rules for this with respect to all manageable bodies: I will mention one, which depends on the property which the centre of gravity has, of always endeavouring to descend to the lowest point.

If a body A (Plate II, Fig. 10) be freely suspended on a pin *a*, and a plumb line *a* B be hung by the same pin, it will pass through the centre of gravity, for that centre is not in the lowest point, till it fall in the same line as the plumb line. Mark the line *a* B; then hang the body up by any other point, as D, with the plumb line D E, which will also pass through the centre of gravity for the same reason as before: and therefore as the centre of gravity is somewhere in *a* B, and also in some point of D E, it must be in the point C, where those lines cross.

CONVERSATION X

Of the Centre of Gravity.

CHARLES. How do those people, who have to load carts and waggons with light goods, as hay, wool, &c., know where to find the centre of gravity?

Father. Perhaps the generality of them never heard of such a principle; and it seems surprising that they should nevertheless make up their loads with such accuracy as to keep the line of direction in or near the middle of the base.

Emma. I have sometimes trembled to pass by the hop-waggons which we have met on the Kent road.

Father. And without any impeachment of your courage, for they are loaded to such an enormous height, that they totter every inch of the road. It would indeed be impossible for one of these to pass with tolerable security along a road much inclined; the centre of gravity being removed so high above the body of the carriage, a small declination on one side or the other would throw the line of direction out of the base.

Emma. When brother James falls about, is it because he cannot keep the centre of gravity between his feet?

Father. That is the precise reason why any person, whether old or young, falls. And hence you learn that a man stands much firmer with his feet a little apart than if they were quite close, for by separating them he increases the base. Hence also the difficulty of sustaining a tall body, as a walking cane, upon a narrow foundation.

Emma. How do rope and wire dancers, whom I have seen at the Circus, manage to balance themselves?

Father. They generally hold a long pole, with weights at each end, across the rope on which they dance, keeping their eyes fixed on some object parallel to the rope, by which means they know when their centre of gravity declines to one side of the rope or the other, and thus, by the help of the pole, they are enabled to keep the centre of gravity over the base, narrow as it is. It is not however rope dancers only that pay attention to this principle, but the most common actions of the people in general are regulated by it.

Charles. In what respects?

Father. We bend forward, when we go up stairs, or rise from our chair, for when we are sitting, our centre of gravity is on the seat, and the line of direction falls behind our base: we therefore lean forwards to bring the line of direction towards our feet. For the same reason a man carrying a burden on his back leans forward; and backward if he carries it on his breast. If the load be placed on one shoulder he leans to the other. If we slip or stumble with one foot, we naturally extend the opposite arm, making the same use of it as the rope-dancer does of his pole.

This property of the centre of gravity always endeavouring to descend will account for appearances, which are sometimes exhibited to excite the surprise of spectators.

Emma. What are those?

Father. One is, that of a double cone, appearing to roll up two inclined planes forming an angle with each other, for as it rolls it sinks between them, and by that means the centre of gravity is actually descending.

Let a body EF (Plate II, Fig. 13), consisting of two equal cones united at their bases, be placed upon the edges of two straight smooth rulers, AB and CD, which at one end meet in an angle at A, and rest on a horizontal plane, and at the other are raised a little above the plane; the body will roll towards the elevated end of the rulers, and appear to ascend; the parts of the cone that rest on the rulers, growing smaller as they go over a large opening, and thus letting it down, the centre of gravity descends. But you must remember that the height of the planes must be less than the radius of the base of the cone.

Charles. Is it upon this principle that a cylinder is made to roll up hill?

Father. Yes, it is; but this can be effected only to a small distance. If a cylinder of pasteboard, or very light wood A B (Plate II, Fig. 11), having its centre of gravity at *c*, be placed on the inclined plane C D, it will roll down the inclined plane, because a line of direction from that centre lies out of the base. If I now fill the little hole *o* above with a plug of lead, it will roll up the inclined plane, till the lead gets near the base, where it will lie still: because the centre of gravity by means of the lead is removed from *c* towards the plug, and therefore is descending, though the cylinder is ascending.

Before I put an end to this subject, I will show you another experiment, which without understanding the principle of the centre of gravity cannot be explained. Upon this stick A (Plate II, Fig. 12), which, of itself, would fall, because its centre of gravity hangs over the table E F, I suspend a bucket B, fixing another stick *a*, one end in a notch between A and *k*, and the other against the inside of the pail at the bottom. Now you will see that the bucket will, in this position, be supported, though filled with water. For the bucket being pushed a little out of the perpendicular, by the stick *a*, the centre of gravity of the whole is brought under the table, and is consequently supported by it.

The knowledge of the principle of the centre of gravity in bodies, will enable you to explain the structure of a variety of toys which are put into the hands of children, such as the *little sawyer; rope dancer; tumbler, &c.*

CONVERSATION XI.

On the Laws of Motion.

CHARLES. Are you now going, papa, to describe those machines, which you call *mechanical powers?*

Father. We must, I believe, defer that a day or two longer, as I have a few more general principles with which I wish you previously to be acquainted.

Emma. What are these?

Father. In the first place, you must well understand what are denominated the three general laws of motion: the *first* of which is, "*that every body will continue in its state of rest or of uniform motion, until it is compelled by some force to change its state.*" This constitutes what is denominated the *inertia* or inactivity of matter. And it may be observed, that a change never happens in the motion of any body, without an equal and opposite change in the motion of some other body.

Charles. There is no difficulty of conceiving that a body, as this inkstand, in a state of rest must always remain so, if no external force be impressed upon it to give it motion. But I know of no example which will lead me to suppose, that a body once put into motion would of itself continue so.

Father. You will, I think, presently admit the latter part of the assertion as well as the former, although it cannot be established by experiment.

Emma. I shall be glad to hear how this is.

Father. You will not deny that the ball which you strike from the trap, has no more power either to destroy its motion, or cause any change in its velocity, than it has to change its shape.

Charles. Certainly; nevertheless, in a few seconds after I have struck the ball with all my force, it falls to the ground, and then stops.

Father. Do you find no difference in the time that is taken up before it comes to rest, even supposing your blow the same?

Charles. Yes, if I am playing on the grass it rolls to a less distance than when I play on the smooth gravel.

Father. You find a like difference when you are playing at marbles, if you play in the gravel court, or on the even pavement in the arcade.

Charles. The marbles run so easily on the smooth stones in the arcade, that we can scarcely shoot with a force small enough.

Emma. And I remember Charles and my cousin were, last winter, trying how far they could shoot their marbles along the ice in the canal; and they went a prodigious distance in comparison of that which they would have gone on the gravel, or even on the pavement in the arcade.

Father. Now these instances properly applied will convince you, that a body once put into motion, would go on for ever, if it were not compelled by some external force to change its state.

Charles. I perceive what you are going to say:—it is the rubbing or friction of the marbles against the ground which does the business. For on the pavement there are fewer obstacles than on the gravel, and fewer on the ice than on the pavement; and hence you would lead us to conclude, that if all obstacles were removed they might proceed on for ever. But what are we to say of the ball, what stops that?

Father. Besides friction, there is another and still more important circumstance to be taken into consideration, which affects the ball, marbles, and every body in motion.

Charles. I understand you, that is the attraction of gravitation.

Father. It is; for from what we said when we conversed on that subject, it appeared that gravity has a tendency to bring every body in motion to the earth; consequently, in a few seconds, your ball must come to the ground by that cause alone; but besides the attraction of gravitation, there is the resistance which the air, through which the ball moves, makes to its passage.

Emma. That cannot be much, I think.

Father. Perhaps, with regard to the ball struck from your brother's trap, it is of no great consideration, because the velocity is but small; but in all great velocities, as that of a ball from a musket or cannon, there will be a material difference between the theory and practice, if it be neglected in the calculation. Move your mamma's riding-whip through the air slowly, and you observe nothing to remind you that there is this resisting medium; but if you swing it with considerable swiftness, the noise which it occasions will inform you of the resistance it meets with from something, which is the atmosphere.

Charles. If I now understand you, the force which compels a body in motion to stop, is of three kinds; (1.) the attraction of gravitation;—(2.) the resistance of the air;—and (3.) the resistance it meets with from friction.

Father. You are quite right.

Charles. I have no difficulty of

conceiving, that a body in motion, will not come to a state of rest, till it is brought to it by an external force, acting upon it in some way or other. I have seen a gentleman, when skaiting on very slippery ice, go a great way without any exertion to himself, but where the ice was rough, he could not go half the distance without making fresh efforts.

Father. I will mention another instance or two on this law of motion. Put a bason of water into your little sister's waggon, and when the water is perfectly still, move the waggon, and the water, resisting the motion of the vessel, will at first rise up in the direction contrary to that in which the vessel moves. If, when the motion of the vessel is communicated to the water, you suddenly stop the waggon, the water, in endeavouring to continue the state of motion, rises up on the opposite side.

In like manner, if while you are sitting quietly on your horse, the animal starts forward, you will be in danger of falling off backward; but if, while you are galloping along, the animal stops on a sudden, you will be liable to be thrown forward.

Charles. This I know by experience, but I was not aware of the reason of it till to-day.

Father. One of the first, and not least important uses of the principles of natural philosophy is, that they may be applied to, and will explain many of the common concerns of life.

We now come to the *second* law of motion, which is;—" *that the change of motion is proportional to the force impressed, and in the direction of that force."*

Charles. There is no difficulty in this, for if while my cricket-ball is rolling along after Henry has struck it, I strike it again, it goes on with increased velocity, and that in proportion to the strength which I exert on the occasion; whereas, if while it is rolling, I strike it back again, or give it a side blow, I change the direction of its course.

Father. In the same way, gravity, and the resistance of the atmosphere, change the direction of a cannon ball from its course in a straight line, and bring it to the ground; and the ball goes to a farther or less distance in proportion to the quantity of powder used.

The *third* law of motion is;—"*that to every action of one body upon another, there is an equal and contrary re-action.*" If I strike this table, I communicate to it (which you perceive by the shaking of the glasses) the motion of my hand: and the table re-acts against my hand, just as much as my hand acts against the table.

If you press with your finger one scale of a balance, to keep it in equilibrio with a pound weight in the other scale, you will perceive, that the scale pressed by the finger acts against it with a force equal to a pound, with which the other scale endeavours to descend.

In all cases the quantity of motion gained by one body is always equal to that lost by the other in the same direction. Thus, if a ball in motion strike another at rest, the motion communicated to the latter will be taken from the former, and the velocity of the former will be proportionally diminished.

A horse drawing a heavy load is as much drawn back by the load as he draws it forward.

Emma. I do not comprehend how the cart draws the horse.

Father. But the progress of the horse is impeded by the load, which is the same thing: for the force which the horse exerts would carry him to a greater distance in the same time, were he freed from the incumbrance of the load, and therefore, as much as his progress falls short of that distance, so much is he, in effect, drawn back by the re-action of the loaded cart.

Again, if you and your brother

were in a boat, and if, by means of a rope, you were to attempt to draw another to you, the boat in which you were would be as much pulled toward the empty boat as that would be moved to you; and if the weights of the two boats were equal, they would meet in a point half way between the two.

If you strike a glass bottle with an iron hammer, the blow will be received by the hammer and the glass; and it is immaterial whether the hammer be moved against the bottle at rest, or the bottle be moved against the hammer at rest, yet the bottle will be broken, though the hammer be not injured, because the same blow, which is sufficient to break glass, is not sufficient to break or injure a mass of iron.

From this law of motion you may learn in what manner a bird, by the stroke of its wings, is able to support the weight of its body.

Charles. Pray explain this, papa.

Father. If the force with which it strikes the air below it is *equal* to the weight of its body, then the re-action of the air upwards is likewise equal to it; and the bird, being acted upon by two *equal* forces in contrary directions, will rest between them. If the force of the stroke is *greater* than its weight, the bird will rise with the *difference* of these two forces: and if the stroke be *less* than its weight, then it will sink with the difference.

CONVERSATION XII.

On the Laws of Motion.

CHARLES. Are those laws of motion which you explained yesterday of great importance in natural philosophy?

Father. Yes, they are, and should be carefully committed to memory. They were assumed by sir Isaac Newton, as the fundamental principles of mechanics, and you will find them at the head of all books written on these subjects. From these also, we are naturally led to some other branches of science, which, though we cannot but slightly mention, should not be wholly neglected. They are, in fact, but corollaries to laws of motion.

Emma. What is a corollary, papa?

Father. It is nothing more than some truth clearly deducible from some other truth before demonstrated or admitted. Thus, by the *first* law of motion, *every body must endeavour to continue in the state into which it is put, whether it be of rest, or uniform motion in a straight line:* from which it follows, as a corollary, "that when we see a body move in a curve line, it must be acted upon by at least two forces."

Charles. When I whirl a stone round in a sling, what are the two forces which act upon the stone?

Father. There is the force, by

which, if you let go the string, the stone will fly off in a right line; and there is the force of the hand, which keeps it in a circular motion.

Emma. Are there any of these circular motions in nature?

Father. The moon, and all the planets move by this law:—to take the moon as an instance. It has a constant tendency to the earth, by the attraction of gravitation, and it has also a tendency to proceed in a right line, by that projectile force impressed upon it by the Creator, in the same manner as the stone flies from your hand; now, by the joint action of these two forces, it describes a circular motion.

Emma. And what would be the consequence, supposing the projectile force to cease?

Father. The moon must fall to the earth; and, if the force of gravity were to cease acting upon the moon, it would fly off into infinite space. Now the projectile force, when applied to the planets, is called the *centrifugal* force, as having a tendency to recede or fly from the centre; and the other force is termed the *centripetal* force, from its tendency to some point as a centre.

Charles. And all this in consequence of the inactivity of matter, by which bodies have a tendency to continue in the same state they are in, whether of rest or motion?

Father. You are right; and this principle which sir Isaac Newton assumed to be in all bodies, he called their *vis inertiæ*, to which we have before referred.

Charles. A few mornings ago, you showed us, that the attraction of the earth upon the moon * is 3600 times less than it is upon heavy bodies near the earth's surface. Now, as this attraction is measured by the space fallen through in a given time, I have endeavoured to calculate the space which the moon would fall through in a minute, were the projectile force to cease.

Father Well, and how have you brought it out ?

Charles. A body falls here 16 feet in the first second, consequently in a minute, or 60 seconds, it would fall 60 times 60 feet, multiplied by 16, that is 3600 feet, which is to be multiplied by 16; and as the moon would fall through 3600 times less space

* See Conversation IV.

in a given time than a body here, it would fall only 16 feet in the first *minute.*

Father. Your calculation is accurate. I will recal to your mind the *second* law, by which it appears, *that every motion or change of motion produced in a body, must be proportional to, and in the direction of, the force impressed.* Therefore, if a moving body receives an impulse in the direction of its motion, its velocity will be increased; if in the contrary direction, its velocity will be diminished;—but if the force be impressed in a direction oblique to that in which it moves, then its direction will be between that of its former motion, and that of the new force impressed.

Charles. This I know from the

observations I have made with my cricket-ball.

Father. By this second law, of motion, you will easily understand, that if a body at rest receives two impulses, at the same time, from forces whose directions do not coincide, it will, by their joint action, be made to move in a line that lies between the direction of the forces impressed.

Emma. Have you any machine to prove this satisfactorily to the senses?

Father. There are many such, invented by different persons, descriptions of which you will hereafter find in various books on these subjects. But it is easily understood by a figure. If on the ball A (Plate II, Fig. 14) a force be impressed sufficient to make it move with an uniform velocity to the point B, in a second of time; and if another force be also impressed on the ball, which alone would make it move to the point C, in the same time; the ball, by means of the two forces, will describe the line A D, which is a diagonal of the figure, whose sides are A C and A B.

Charles. How then is motion produced in the *direction of the force:* according to the second law, it ought to be, in one case, in the direction A C, and, in the other, in that of A B, whereas it is in that of A D.

Father. Examine the figure a little attentively, carrying this in your mind, that, for a body to move in the *same direction*, it is *not* necessary that it should move in the *same straight line*; but that it is sufficient to move *either* in that line, or in any one parallel to it.

Charles. I perceive then that the ball, when arrived at D, has moved in the direction A C, because B D is parallel to A C; and also in the direction A B, because C D is parallel to it.

Father. And in no other possible situation but at the point D, could this experiment be conformable to the second law of motion.

You will not forget, that, if a body move in a curve, the *continued* action of external force must be inferred: if that action were to cease at any point the body would continue its motion in a straight line touching the curvilineal path in that point.

CONVERSATION XIII.

Of the Laws of Motion.

FATHER. If you reflect a little upon what we said yesterday on the second law of motion, you will readily deduce the following corollaries. (Plate II, Fig. 14.)

1. That, if the forces be equal, and act at right angles to one another, the line described by the ball will be the diagonal of a *square*. But in all other cases it will be the diagonal of a parallelogram of some kind.

2. By varying the angle and the forces, you vary the form of your parallelogram.

Charles. Yes, papa; and I see another consequence, *viz.* that the motions of two forces acting conjointly in this way are not so great as when they act separately.

Father. That is true; and you are led to the conclusion, I suppose, from the recollection, that in every triangle any two sides taken together are greater than the remaining side; and therefore you infer, and justly too, that the motions which the ball A must have received, had the forces been applied separately, would have been equal to A C and A B, or, which is the same thing, to A C and C D, the two sides of the triangle A D C; but by their joint action the motion is only equal to A D, the remaining side of the triangle.

Hence then you will remember,

that in the *composition*, or adding together of forces (as this is called), motion is always lost: and in the *resolution* of any one force, as A D, into two others, A C and A B, motion is gained.

Charles. Well, papa, but how is it that the heavenly bodies, the moon for instance, which is impelled by two forces, performs her motion in a circular curve round the earth, and not in a diagonal between the direction of the projectile force and that of the attraction of gravity to the earth?

Father. Because, in the case just mentioned, there was only the action of a single impulse in each direction, whereas the action of gravity on the moon is continual, and causes an accelerated motion, and hence the line is a curve.

Charles. Supposing then, that A represent the moon, and A C the sixteen feet through which it would fall in a second by the attraction of gravity towards the earth, and A B represent the projectile force acting upon it for the same time. If A B and A C acted as single impulses, the moon would in that case describe the diagonal A D; but since these forces are constantly acting, and that of gravity is an accelerating force also, therefore instead of the straight line A D, the moon will be drawn into the curve line A *a* D. Do I understand the matter right?

Father. You do; and hence you easily comprehend how, by good instruments and calculation, the attraction of the earth upon the moon was discovered.

The *third* law of motion, viz. *that action and re-action are equal and in contrary directions*, may be illustrated by the motion communicated by the percussion of *elastic* and *non-elastic* bodies.

Emma. What are these, papa?

Father. *Elastic* bodies are those which have a certain spring, by which their parts, upon being pressed inwards, by percussion, return to their former state; this property is evident in a ball of wool or cotton, or in sponge compressed. *Non-elastic* bodies are those which, when one strikes another, do not rebound, but move together after the stroke.

Let two *equal* ivory balls *a* and *b* be suspended by threads; if *a* (Plate II. Fig. 15) be drawn a little out of the perpendicular, and let fall upon *b*, it will lose its motion by communicating it to *b*, which will be driven to a distance *c*, equal to that through which *a* fell; and hence it appears that the re-action of *b* was equal to the action of *a* upon it.

Emma. But do the parts of the ivory balls yield by the stroke, or, as you call it, by percussion?

Father. They do; for if I lay a little paint on *a*, and let it *touch b*, it will make but a very small speck upon it: but if it *fall* upon *b*, the speck will be much larger; which proves that the balls are elastic, and that a little hollow, or dint, was made in each by collision. If now two equal soft balls of clay, or glazier's putty, which are non-elastic, meet each other with equal velocities, they would stop and stick together at the place of their meeting, as their mutual actions destroy each other.

Charles. I have sometimes shot my white alley against another marble so plumply, that the marble has gone off as swiftly as the alley approached it, but the alley remained motionless in the place of the marble. Are marbles, therefore, as well as ivory, elastic?

Father. They are.—If three elastic balls, *a*, *b*, *c* (Plate III, Fig. 16), be hung from adjoining centres, and *c* be drawn a little out of the perpendicular, and let fall upon *b*, then will *c* and *b* become stationary, and *a* will be driven to *o*, the distance through which *c* fell upon *b*.

If you hang any number of balls, as six, eight, &c., so as to touch each other, and if you draw the outside one away to a little distance, and then let it fall upon the others, the ball on the opposite side will be driven off while the rest remain stationary, so equal is the action and re-action of the stationary balls divided among them. In the same manner, if two are drawn aside and suffered to fall on the rest, the opposite two will fly off, and the others remain stationary.

There is one other circumstance depending upon the action and re-action of bodies, and also upon the *vis inertiæ* of matter, worth noticing: by some authors you will find it largely treated upon.

If I strike a blacksmith's anvil with a hammer, action and re-action being equal, the anvil strikes the hammer as forcibly as the hammer strikes the anvil.

If the anvil be large enough, I might lay it on my breast, and suffer you to strike it with a sledge hammer with all your strength, without pain or risque, for the *vis inertiæ* of the anvil resists the force of the blow. But if the anvil were but a pound or two in weight, your blow would probably kill me.

Emma. Is it owing to this principle, that when a cannon on wheels is fired, it runs backward?

Father. It is; for the action of the powder produces as great a quantity of motion in the gun, as in the ball, but their motions are contrary; the ball moves forward and the cannon backward.

CONVERSATION XIV.

On the Mechanical Powers.

CHARLES. Will you now, papa, explain the mechanical powers?

Father. I will, and I hope you have not forgotten what the *momentum* of a body is?

Charles. No; it is the force of a moving body, which force is estimated by the weight, multiplied into its velocity.

Father. Then a small body may have an equal momentum with one much larger?

Charles. Yes, provided the smaller body moves as much swifter than the larger one, as the weight of the latter is greater than that of the former.

Father. What do you mean when you say, that one body moves swifter, or has a greater velocity than another?

Charles. That it passes over a greater space in the same time. Your watch will explain my meaning: the minute-hand travels round the dial-plate in an hour, but the hour-hand takes twelve hours to perform its course in, consequently the velocity of the minute-hand is twelve times greater than that of the hour-hand; because, in the same time, *viz.* twelve hours, it travels twelve times the space that is gone through by the hour-hand.

Father. But this can be only

true on the supposition, that the two circles are equal. In my watch, the minute hand is longer than the other, and, consequently, the circle described by it is larger than that described by the hour-hand.

Charles. I see at once, that my reasoning holds good only in the case where the hands are equal.

Father. There is, however, a particular point of the longer hand, of which it may be said, with the strictest truth, that it has exactly twelve times the velocity of the extremity of the shorter.

Charles. That is the point, at which, if the remainder were cut off, the two hands would be equal. And, in fact, every different point of the hand, describes different spaces in the same time.

Father. The little pivot on which the two hands seem to move (for they are really moved by different pivots, one within another) may be called the *centre of motion*, which is a fixed point; and the longer the hand is, the greater is the space described.

Charles. The extremities of the vanes of a wind-mill, when they are going very fast, are scarcely distinguishable, though the separate parts, nearer the mill, are easily discerned; this is owing to the velocity of the extremities being so much greater than that of the other parts.

Emma. Did not the swiftness of the round-abouts, which we saw at the fair, depend on the same principle, *viz.* the length of the poles upon which the seats were fixed?

Father. Yes, the greater the distance at which these seats were placed from the centre of motion, the greater was the space which the little boys and girls travelled for their halfpenny.

Emma. Then those in the second row had a shorter ride for their money, than those at the end of the poles?

Father. Yes, shorter as to space, but the same as to time. In the same way, when you and Charles go round the gravel-walk for half an hour's exercise, if he run while you walk, he will, perhaps, have gone six or eight times round, in the same time that you have been but three or four times; now, as to time, your exercise has been equal, but he may have passed over double the space in the same time.

Charles. How does this apply to the explanation of the mechanical powers?

Father. You will find the application very easy:——without clear ideas of what is meant by *time* and *space*, you cannot comprehend the principles of mechanics.

There are six mechanical powers: the lever; the wheel and axis; the pulley; the inclined plane; the wedge; and the screw.

Emma. Why are they called mechanical powers?

Father. Because, by their means, we are enabled *mechanically* to raise weights, move heavy bodies, and overcome resistances, which, without their assistance, could not be done.

Charles. But is there no limit to the assistance gained by these powers? for I remember reading of Archimedes, who said, that with a place for his fulcrum he could move the earth itself.

Father. Human power, with all the assistance which art can give, is very soon limited, and upon this principle, *that what we gain in power, we lose in time.* That is, if by your own unassisted strength, you are able to raise fifty pounds to a certain distance in one minute, and if, by the help of machinery, you wish to raise 500 pounds to the same height, you will require ten minutes to perform it in: thus you increase your power ten-fold, but it is at the expense of time. Or, in other words, you are enabled to do that with one effort in ten minutes, which you could have done in ten separate efforts in the same time.

Emma. The importance of mechanics, then, is not so very considerable as one, at first sight, would imagine: since there is no real gain of force acquired by the mechanical powers.

Father. Though there be not any actual increase of force gained by these powers; yet the advantages which men derive from them are inestimable. If there are several small weights, manageable by human strength, to be raised to a certain height, it may be full as convenient to elevate them one by one, as to take the advantage of the mechanical powers in raising them all at once. Because, as we have shown, the same time will be necessary in both cases. But suppose you have a large block of stone of a ton weight to carry away, or a weight still greater, what is to be done?

Emma. I did not think of that.

Father. Bodies of this kind cannot be separated into parts proportionable to the human strength without immense labour, nor, perhaps, without rendering them unfit for those purposes to which they are to be applied. Hence then you perceive the great importance of the mechanical powers, by the use of which a man is able with ease to manage a weight many times greater than himself.

Charles. I have, indeed, seen a

few men, by means of pulleys, and apparently with no very great exertion, raise an enormous oak tree into a timber-carriage, in order to convey it to the dock-yard.

Father. A very excellent instance; for if the tree had been cut into such pieces as could have been managed by the natural strength of these men, it would not have been worth carrying to Deptford or Chatham for the purpose of ship building.

Emma. I acknowledge my error; —what is a fulcrum?

Father. It is a *fixed point*, or prop, round which the other parts of a machine move.

Charles. Is the pivot, upon which the hands of your watch move, a fulcrum, then?

Father. It is, and you remember

we called it also the centre of motion: the rivet of these scissars is also a fulcrum, and also the centre of motion.

Emma. Is that a fixed point or prop?

Father. Certainly it is a fixed point, as it regards the two parts of the scissars; for that always remains in the same position, while the other parts move about it. Take the poker and stir the fire:—now that part of the bar on which the poker rests is a fulcrum, for the poker moves upon it as a centre.

CONVERSATION XV.

Of the Lever.

FATHER. We will now consider the *Lever*, which is generally called the first mechanical power.

The *lever* is an inflexible bar of wood, iron, &c., which serves to raise weights, while it is supported at a point by a prop or fulcrum, on which, as the centre of motion, all the other parts turn. A B (Plate III, Fig. 17) will represent a lever, and the point C the fulcrum or centre of motion. Now, it is evident, if the lever turn on its centre of motion C, so that A comes into the position *a*; B at the same time must come into the position *b*. If both the arms of the lever be equal, that is, if A C is equal to B C, there is no advantage gained by it, for they pass over equal spaces in the same time; and, according to the fundamental principles already laid down (p. 127, "as advantage or power is gained, time must be lost:" therefore, no time being lost by a lever of this kind, there can be no power gained.

Charles. Why then is it called a mechanical power?

Father. Strictly speaking perhaps it ought not to be numbered as one. But it is usually reckoned among them, having the fulcrum between the weight and the power, which is the distinguishing property of levers of the first kind. And, when the fulcrum is exactly the middle point between the weight and power, it is the common balance: to which, if scales be suspended at A and B, it is fitted for weighing all sorts of commodities.

Emma. You say it is a lever of the *first* kind, are there several sorts of levers?

Father. There are three sorts; some persons reckon four; the fourth, however, is but a bended one of the first kind. A lever of the *first* kind (Plate III, Fig. 18, 19) has the fulcrum between the weight and power.

The *second* kind of lever (Plate III, Fig. 20) has the fulcrum at one end, the power at the other, and the weight between them.

In the *third* kind (Plate III, Fig. 21) the power is between the fulcrum and the weight.

Of Levers' powers the different sorts are three,
The *first* in steel-yards and in scales you see;
The best, a *second*, is the miller's lift,
Where *power* and *fulcrum* to each end you
shift;
And in the *third*, the worst of all, my friend,
You find the *weight* and *fulcrum* at each end.

Let us take the lever of the first kind (Fig. 18), which, if it be moved into the position *a b*, by turning on its fulcrum C, it is evident that while A has travelled over the short space A *a*, B has travelled over the greater space B *b*, which spaces are to one another, exactly in proportion to the length of the arms A C and B C. If now you apply your hand first to the point A, and afterwards to B, in order to move the lever into the position *a b*, using the same velocity in both cases, you will find, that the time spent in moving the lever when the hand is at B, will be as much greater as that spent when the hand is at A, as the arm B C is longer than the arm A C; but then the exertion required will, in the same proportion, be less at B than at A.

Charles. The arm B C appears to be four times the length of A C.

Father. Then it is a lever which gains power in proportion of four to one. That is, a single pound weight applied to the end of the arm B C, as at P, will balance four pounds suspended at A, as W.

Charles. I have seen workmen move large pieces of timber to very small distances, by means of a long bar of wood or iron; is that a lever?

Father. It is; they force one end of the bar under the timber, and then place a block of wood, stone, &c., beneath, and as near the same end of the lever as possible, for a fulcrum, applying their own strength to the other: and power is gained in proportion as the distance from the fulcrum to the part where the men apply their strength, is greater than the distance from the fulcrum to that end under the timber. Handspikes are levers of this kind, and by these the heaviest cannon are moved.

Charles. It must be very considerable, for I have seen two or three men move a tree, in this way, of several tons weight, I should think.

Father. That is not difficult; for supposing a lever to gain the advantage of twenty to one, and a man by his natural strength is able to move but a hundred weight, he will find that by a lever of this sort, he can move twenty hundred weight, or a ton; but for single exertions, a strong man can put forth a much greater power, than that which is sufficient to remove a hundred weight; and levers are also frequently used, the advantage gained by which is still more considerable than twenty to one.

There is another method, by means of a lever of the first kind, of moving, and even of pulling trees up by their roots. A strong scantling of timber is fixed perpendicularly to the axle of a pair of cart wheels, which is strapped firmly to the tree; and when the lateral roots are cut, by digging a trench round it, the tap root or roots are easily torn up by a team of two or three horses; for the tree itself becomes the lever, and the axle of the wheels its fulcrum.

Charles. I think you said, the other day, that the common steelyard made use of by the butcher, is a lever?

Father. I did; the short arm A C (Plate III, Fig 19) is, by an increase in size, made to balance the longer one B C, and from C, the centre of motion, the divisions must commence. Now if B C be divided into as many parts as it will contain, each equal to A C, a single weight, as a pound P, will serve for weighing any thing as heavy as itself, or as many times heavier as there are divisions in the arm C. If the weight P be placed at the division 1, in the arm B C, it will balance one pound in the scale at A; if it be removed to 3, 5, or 7, it will balance 3, 5, or 7 pounds in the scale; for these divisions being respectively 3, 5, or 7 times the distance from the centre of motion C that A is, it becomes a lever, which gains advantage at those points, in the proportion of 3, 5, and 7. If now the intervals between the divisions on the longer arm be subdivided into halves, quarters, &c., any weight may be accurately ascertained to halves, quarters of pounds, &c.

CONVERSATION XVI.

Of the Lever.

EMMA. What advantage has the steel-yard, which you described in our last Conversation, over a pair of scales?

Father. It may be much more readily removed from place to place; it requires no apparatus, and only a single weight for all the purposes to which it can be applied. Sometimes the arms are not of equal weight: in that case the weight P must be moved along the arm BC, till it exactly balance the other arm without a weight; and in that point a notch must be made, marking over it a cipher 0, from whence the divisions must commence.

Charles. Does there require great accuracy in the manufacture of instruments of this kind?

Father. Yes; of such importance is it to the public, that there should be no error or fraud by means of false weights, or false balances, that it is the business of certain public officers to examine at stated seasons the weights, measures, &c., of every shopkeeper in the land. Yet it is to be feared, that, after all precautions, much fraud is practised upon the unsuspecting.

Emma. I one day last summer bought, as I supposed, a pound of cherries at the door, but Charles thinking there were not a pound we tried them in our scales, and found but twelve ounces, or three quarters, instead of a pound, and yet the scale went down as if the man had given me full weight. How was that managed?

Father. It might be done many ways: by short weights;—or by the scale in which the fruit was put being heavier than the other;—but fraud may be practised with honest weights and scales, by making the arm of the balance on which the weight hangs shorter than the other, for then a pound weight will be balanced by as much less fruit than a pound, as that arm is shorter than the other; this was probably the method by which you were cheated.

Emma. By what method could I have discovered this cheat?

Father. The scales when empty are exactly balanced, but when loaded, though still in equilibrio, the weights are unequal, and the deceit is instantly discovered by changing the weights to the contrary scales. I will give you a rule to find the true weight of any body by such a false balance, the reason of the rule you will understand hereafter: "*Find the weights of the body by both scales, multiply them together, and then find the square root of the product, which is the true weight.*"

Charles. Let me see if I understand the rule: suppose a body weigh 16 ounces in one scale, and in the other 12 ounces and a quarter, I multiply 16 by 12 and a quarter, and

I get the product 196, the square root of which is 14: for 14 multiplied into itself gives 196; therefore the true weight of the body is 14 ounces.

Father. That is just what I meant. —To the lever of the first kind may be referred many common instruments, such as scissars, pincers, snuffers, &c., which are made by two levers, acting contrary to one another.

Emma. The rivet is the fulcrum, or centre of motion, the hand the power used, and whatever is to be cut is the resistance to be overcome.

Charles. A poker stirring the fire is also a lever, for the bar is the fulcrum, the hand the power, and the coals the resistance to be overcome.

Father. We now proceed to levers of the second kind, in which the fulcrum c (Fig 20) is at one end, the power P applied at the other B, and the weight to be raised W, somewhere between the fulcrum and the power.

Charles. And how is the advantage gained to be estimated in this lever?

Father. By looking at the figure you will find that power or advantage is gained in proportion as the distance B, the point at which the power P acts, is greater than the distance of the weight W from the fulcrum.

Charles. Then if the weight hang at one inch from the fulcrum, and the power acts at five inches from it, the power gained is five to one, or one pound at P will balance five at W?

Father. It will: for you perceive that the power passes over five times as great a space as the weight, or while the point A in the lever moves over one inch, the point B will move over five inches.

Emma. What things in common use are to be referred to the lever of the second kind?

Father. The most common and useful of all things; every door, for instance, which turns on hinges, is a lever of this sort. The hinges may be considered as the fulcrum or centre of motion, the whole door is the weight to be moved, and the power is applied to that side on which the lock is usually fixed.

Emma. Now I see the reason why there is considerable difficulty in pushing open a heavy door, if the hand is applied to the part next the hinges, although it may be opened with the greatest ease in the usual method.

Charles. This sofa, with sister

upon it, represents a lever of the second kind?

Father. Certainly, if, while she is sitting upon it in the middle, you raise one end, while the other remains fixed as a prop or fulcrum. To this kind of lever may be also reduced nut-crackers; oars; rudders of ships; those cutting-knives which have one end fixed in a block, such as are used for cutting chaff, drugs, wood for pattens, &c.

Emma. I do not see how oars and rudders are levers of this sort.

Father. The boat is the weight to be moved, the water is the fulcrum, and the waterman at the handle the power. The masts of ships are also levers of the second kind, for the bottom of the vessel is the fulcrum, the ship the weight, and the wind acting against the sail is the moving power.

The knowledge of this principle may be useful in many situations and circumstances of life:—if two men unequal in strength have a heavy burden to carry on a pole between them, the ability of each may be consulted by placing the burden as much nearer to the stronger man, as his strength is greater than that of his partner.

Emma. Which would you call the prop in this case?

Father. The stronger man, for the weight is nearest to him, and then the weaker must be considered as the power. Again, two horses may be so yoked to a carriage that each shall draw a part proportional to his strength, by dividing the beam in such a manner, that the point of *traction*, or drawing, may be as much nearer to the stronger horse than to the weaker, as the strength of the former exceeds that of the latter.

The principle of the wheel-barrow may be referred to a lever of the second kind. The fulcrum C (Fig. 20) may be considered as the wheel, W the load, and B the place where the hands are applied: hence a man is enabled to drive or drag a much heavier load than he could carry, because his *power* at B is applied farther from the centre of motion C than the weight W.

We will now describe the third kind of lever. In this the prop or fulcrum C (Fig. 21) is at one end, the weight W at the other, and the power P is applied at B, somewhere between the prop and weight.

Charles. In this case, the weight being farther from the centre of motion than the power, must pass through more space than it.

Father. And what is the consequence of that?

Charles. That the power must be greater than the weight, and as much greater as the distance of the weight from the prop exceeds the distance of the power from it, that is, to balance a weight of three pounds at A, there will require the exertion of a power P, acting at B, equal to five pounds.

Father. Since then a lever of this kind is a disadvantage to the moving power, it is but seldom used, and only in cases of necessity; such as in that of a ladder, which, being fixed at one end against a wall or other obstacle, is, by the strength of a man's arm, raised into a perpendicular situation. But the most important application of this third kind of lever is manifest in the structure of the limbs of animals, particularly in those of man: to take the arm as an instance; when we lift a weight by the hand, it is effected by means of muscles coming from the shoulder blade, and terminating about one tenth as far below the elbow as the hand is: now the elbow being the centre of motion round which the lower part of the arm turns, according to the principle just laid down, the muscles must exert a force ten times as great as the weight that is raised. At first view, this may appear a disadvantage, but what is lost in power is gained in velocity, and thus the human figure is better adapted to the various functions it has to perform.

CONVERSATION XVII.

Of the Wheel and Axis.

FATHER. Well, Emma, do you understand the principle of the lever, which we discussed so much at large yesterday?

Emma. The lever gains advantage in proportion to space passed through by the acting power; that is, if the weight to be raised be at the distance of one inch from the fulcrum, and the power is applied nine inches distant from it, then it is a lever, which gains advantage as 9 to 1, because the space passed through by the *power* is nine times greater than that passed through by the weight; and, therefore, what is lost in time by passing through a greater space, is gained in power.

Father. You recollect, also, what the different kinds of levers are, I hope.

Emma. I shall never see the fire stirred without thinking of a simple lever of the first kind; my scissars will frequently remind me of a combination of two levers of the same sort. The opening and shutting of the door will prevent me from forgetting the nature of the lever of the second kind: and I am sure, that I shall never see a workman raise a ladder against a house, without recollecting the third sort of lever. Besides, I believe a pair of tongs is a lever of this kind.

Father. You are right; for the fulcrum is at the joint, and the power is applied between that and the parts used in taking up coals, &c.—Can you, Charles, tell us how the principle of *momentum* applies to the lever?

Charles. The *momentum* of a body is estimated by its weight, multiplied into its velocity; and the velocity must be calculated by the space passed through in a given time. Now, if I examine the lever (Fig. 18, 20), and consider it as an inflexible bar, turning on a centre of motion, it is evident, that the same time is used for the motion both of the weight and the power, but the spaces passed over are very different; that which the power passes through being as much greater than that passed by the weight, as the length of the distance of the power from the prop is greater than the distance of the weight from the prop; and the velocities, being as the spaces passed in the same time, must be greater in the same proportion. Consequently, the velocity of P, the power, multiplied into its weight, will be equal to the smaller velocity of W, multiplied into its weight, and thus, their momenta being equal, they will balance one another.

Father. This applies to the first and second kind of lever; what do you say to the third?

Charles. In the third, the velocity of the power P (Fig. 21), being less than that of the weight W, it is evident, in order that their momenta may be equal, that the weight acting at P must be as much greater than that of W as A C is less than B C, and then they will be in equilibrio.

Father. The second mechanical power is the *Wheel and Axis,* which gains power in proportion as the circumference of the wheel is greater than that of the axis; this machine may be referred to the principle of the lever. A B (Plate III, Fig 22) is the wheel, Q D its axis; and, if the circumference of the wheel be eight times as great as that of the axis, then a single pound P will balance a weight W of eight pounds.

Charles. Is it by an instrument of this kind that water is drawn from those deep wells so common in many parts of the country?

Father. It is; but as in most

cases of this kind only a single bucket is raised at once, there requires but little power in the operation, and therefore, instead of a large wheel as AB, an iron handle fixed at a is made use of, which, you know, by its circular motion, answers the purpose of a wheel.

Charles. I once raised some water by a machine of this kind, and I found that as the bucket ascended nearer the top the difficulty increased.

Father. That must always be the case, where the wells are so deep as to cause, in the ascent, the rope to coil more than once the length of the axis, because the advantage gained is in proportion as the circumference of the wheel is greater than that of the axis; so that if the circumference of the wheel be 12 times greater than that of the axis, 1 pound applied at the former will balance 12 hanging at the latter; but by the coiling of the rope round the axis, the *difference* between the circumference of the wheel and that of the axis continually diminishes, consequently the advantage gained is less every time a new coil of rope is wound on the whole length of the axis; this explains why the difficulty of drawing the water, or any other weight, increases as it ascends nearer the top.

Charles. Then by diminishing the axis, or by increasing the length of the handle, advantage is gained?

Father. Yes, by either of those methods we may gain power; but it is very evident that the axis cannot be diminished beyond a certain limit, without rendering it too weak to sustain the weight; nor can the handle be managed if it be constructed on a scale much larger than what is commonly used.

Charles. We must then have recourse to the wheel with spikes standing out of it at certain distances from each other to serve as levers.

Father. You may by this means increase your power according to your wish, but it must be at the expense of time, for you know that a simple handle may be turned several times while you are pulling the wheel round once. To the principle of the *wheel and axis* may be referred the capstan, windlass, and all those numerous kinds of cranes, which are to be seen at the different wharfs on the banks of the Thames.

Rous'd from repose aloft the sailors swarm,
And with their *levers* soon the *windlass* arm.
The order given, up-springing with a bound
They lodge the bars and wheel their engines round:
At ev'ry turn the clanging pauls resound.
Upturn, reluctant, from its oozy cave,
The pond'rous anchor rises o'er the wave.

FALCONER'S SHIPWRECK.

Charles. I have seen a crane, which consists of a wheel large enough for a man to walk in.

Father. In this the weight of the man, or men (for there are sometimes two or three), is the moving power; for, as the man steps forwards, the part upon which he treads becomes the heaviest, and consequently descends till it be the lowest.

On the same principle, you may see at the door of many bird-cage makers, a bird, by its weight, give a wicker cage a circular motion; now, if there were a small weight suspended to the axis of the cage, the bird by its motion would draw it up, for as it hops from the bottom bar to the next, its *momentum* causes that to descend, and thus the operation is performed, both with regard to the cage, and to those large cranes which you have seen.

Emma. Is there no danger if the man happen to slip?

Father. If the weight be very great, a slip with the foot may be attended with very dangerous consequences. To prevent which, there is generally fixed at one end of the axis a little wheel G (Fig. 22), called a rachet-wheel; with a catch H, to fall into its teeth; this will at any time support the weight in case of an accident. Sometimes, instead of men walking within the great wheel, cogs are set round it on the outside, and a small trundle wheel made to work in the cogs, and to be turned by a winch.

Charles. Are there not other sorts of cranes, in which all danger is avoided?

Father. The crane is a machine of such importance to the commercial concerns of this country, that new inventions of it are continually offered to the public: I will, when we go to the library, show you, in the tenth volume of the Transactions of the Society for the Encouragement of Arts and Sciences, an engraving of a safe, and, I believe, truly excellent crane; it was invented by a friend of mine, Mr. James White, who possessed a most extraordinary genius for mechanics, and who formerly offered his services to a noble duke, then at the head of the Board of Ordnance; but these being rejected, he went to the Continent, where, for some years, he was very profitably exercising his talents. But he is now settled as a manufacturer in his own country.

Charles. You said that this mechanical power might be considered as a lever of the first kind.

Father. I did; and if you conceive the wheel and axis (Fig. 22) to be cut through the middle in the direction A B; F G B (Plate III, Fig. 23) will represent a section of it. A B is a lever, whose centre of motion is C; the weight W, sustained by the rope A W, is applied at the distance C A, the radius of the axis; and the power P, acting in the direction B P, is applied at the distance C B, the radius of the wheel; therefore, according to the principle of the lever, the power will balance the weight when it is as much less than the weight, as the distance C B is greater than the distance of the weight A C.

CONVERSATION XVIII.

Of the Pulley.

FATHER. The third mechanical power, the *pulley*, may be likewise explained on the principle of the lever. The line A B (Plate IV, Fig. 24) may be conceived to be a lever, whose arms A C and B C are equal, and C the fulcrum, or centre of motion. If now two equal weights, W and P, be hung on the cord passing over the pulley, they will balance one another, and the fulcrum will sustain both.

Charles. Does this pulley then, like the common balance, give no advantage?

Father. From the single *fixed* pulley no mechanical advantage is derived; it is, nevertheless, of great importance in changing the direction of a power, and is very much used in buildings for drawing up small weights, it being much easier for a man to raise such burdens by means of a single pulley, than to carry them up a long ladder.

Emma. Why is it called a mechanical power?

Father. Though a single fixed pulley gives no advantage, yet when it is not fixed, or when two or more are combined into what is called a system of pulleys, they then possess all the properties of the other mechanical powers. Thus in C D B (Plate IV, Fig. 25) C is the fulcrum, therefore a power P, acting at B, will sustain a double weight W, acting at A, for B C is double the distance of A C from the fulcrum.

Again, it is evident, in the present case, that the whole weight is sustained by the cord E D P, and whatever sustains half the cord, sustains also half the weight; but one half is sustained by the fixed hook E, consequently the power at P has only the other half to sustain, or, in other words, any given power at P will keep in equilibrio a double weight at W.

Charles. Is the velocity of P double that of W?

Father. Undoubtedly; if you compare the space passed through by the hand at P with that passed by W, you will find that the former is just double of the latter, and therefore the *momenta* of the power and weight, as in the lever, are equal.

Charles. I think I see the reason of this, for if the weight be raised an inch, or a foot, both sides of the cord must also be raised an inch, or foot, but this cannot happen without that part of the cord at P passing through two inches, or two feet of space.

Father. You will now easily infer, from what has been already shown of the single *moveable* pulley, that, in a system of pulleys, the power gained must be estimated, by doubling the number of pulleys in the lower or moveable block. So that when the fixed block *x* (Plate IV,

Fig. 26) contains two pulleys which only turn on their axes, and the lower block *y* contains also two pulleys, which not only turn on their axes, but also *rise* with the weight, the advantage is as four; that is a single pound at P will sustain four at W.

Charles. In the present instance also I perceive, that by raising W an inch, there are four ropes shortened each an inch, and therefore the hand must have passed through four inches of space in raising the weight a single inch; which establishes the maxim, that what is gained in power is lost in space. But you have only talked of the power of balancing or sustaining the weight, something more must, I suppose, be added to raise it.

Father. There must; considerable allowance must likewise be made for the friction of the cords, and of the pivots, or axes, on which the pulleys turn. In the mechanical powers, in general, one third of the power must be added for the loss sustained by friction, and for the imperfect manner in which machines are commonly constructed. Thus, if by *theory* you gain a power of 600; in *practice*, you must reckon only upon 400. In those pulleys which we have been describing, writers have noticed three things, which take much from the general advantage and convenience of pulleys as a mechanical power. The *first* is, that the diameters of the axes bear a great proportion to their own diameters. The *second* is, that in working they are apt to rub against one another, or against the side of the block. And the *third* disadvantage is the stiffness of the rope that goes over and under them.

The first two objections have been, in a great degree, removed by the concentric pulley, invented by Mr. James White: B (Plate IV, Fig. 27) is a solid block of brass, in which grooves are cut, in the proportion of 1, 3, 5, 7, 9, &c.; and A is another block of the same kind, whose grooves are in the proportion of 2, 4, 6, 8, 10, &c., and round these grooves a cord is passed, by which means they answer the purpose of so many distinct pulleys, every point of which moving with the velocity of the string in contact with it, the whole friction is removed to the two centres of motion of the blocks A and B; besides it is of no small advantage, that the pulleys being all of one piece, there is no rubbing one against the others.

Emma. Do you calculate the power gained by this pulley in the same method as with the common pulleys?

Father. Yes, for pulleys of every kind the rule is general, the advantage gained is found by doubling the number of the pulleys in the lower block: in that before you there are six grooves, which answer to as many distinct pulleys, and consequently the power gained is twelve, or one pound at P will balance twelve pounds at W.

CONVERSATION XIX.

Of the Inclined Plane.

FATHER. We may now describe the inclined plane, which is the fourth mechanical power.

Charles. You will not be able, I think, to reduce this also to the principle of the lever.

Father. No, it is a distinct principle, and some writers on these subjects reduce at once the six mechanical powers to two, *viz.* the lever and inclined plane.

Emma. How do you estimate the

advantage gained by this mechanical power?

Father. The method is very easy, for just as much as the length of the plane exceeds its perpendicular height, so much is the advantage gained. Suppose A B (Plate IV, Fig. 28) is a plane standing on the table, and C D another plane inclined to it; if the length C D be three times greater than the perpendicular height; then the cylinder E will be supported upon the plane C D, by a weight equal to the third part of its own weight.

Emma. Could I then draw up a weight on such a plane with a third part of the strength that I must exert in lifting it up at the end?

Father. Certainly you might;

allowance, however, must be made for overcoming the friction; but then you perceive, as in other mechanical powers, that you will have three times the space to pass over, or that as you gain power you will lose time.

Charles. Now I understand the reason why sometimes there are two or three strong planks laid from the street to the ground-floor warehouses, making therewith an inclined plane, on which heavy packages are raised or lowered.

Father. The inclined plane is chiefly used for raising heavy weights to small heights; for in warehouses situated in the upper part of buildings, cranes and pulleys are better adapted for the purpose.

Charles. I have sometimes amused myself by observing the difference of time which one marble has taken to roll down a smooth board, and another which has fallen by its own gravity without any support.

Father. And if it were a long plank, and you took care to let both marbles drop from the hand at the same instant, I dare say you found the difference very evident.

Charles. I did, and now you have enabled me to account for it very satisfactorily, by showing me that as much more time is spent in raising a body along an inclined plane, than in lifting it up at the end, as that plane is longer than its perpendicular height. For I take it for granted that the rule holds in the descent as well as in the ascent.

Father. If you have any doubt remaining, a few words will make every thing clear. Suppose your marbles placed on a plane perfectly horizontal, as on this table, they will remain at rest wherever they are placed: now if you elevated the plane in such a manner that its height should be equal to half the length of the plane, it is evident from what has been shown before, that the marbles would require a force equal to half their weight to sustain them in any particular position: suppose then the plane perpendicular to the table, the marbles will descend with their whole weight, for now the plane contributes in no respect to support them, consequently they would require a power equal to their whole weight to keep them from descending.

Charles. And the swiftness with which a body falls is to be estimated by the force with which it was acted upon?

Father. Certainly, for you are now sufficiently acquainted with philosophy to know that the effect must be estimated from the cause. Suppose an inclined plane is thirty-two feet long, and its perpendicular height is sixteen feet, what time will a marble take in falling down the plane, and also in descending from the top to the earth by the force of gravity?

Charles. By the attraction of gravitation, a body falls sixteen feet in a second (see p. 61), therefore the marble will be one second in falling perpendicularly to the ground; and as the length of the plane is double its height, the marble must take two seconds to roll down it.

Father. I will try you with another example. If there be a plane 64 feet perpendicular height, and 3 times 64, or 192 feet long, tell me what time a marble will take in falling to the earth by the attraction of gravity, and how long it will be in descending down the plane.

Charles. By the attraction of gravity it will fall in two seconds; because, by multiplying the sixteen feet which it falls in the first second, by the square of two seconds (the time) or four, I get sixty-four, the height of the plane. But the plane being three times as long as it is perpendicularly high, it must be three times as many seconds in rolling down the plane as it was in descending freely by the force of gravity, that is, six seconds.

Emma. Pray what common instruments are to be referred to this mechanical power, in the same way as scissars, pincers, &c., are referred to the lever?

Father. Chisels, hatchets, and whatever other sharp instruments which are chamfered, or sloped down to an edge on one side only, may be referred to the principle of the inclined plane.

The principle of the inclined plane is applied in the construction of carriage-ways, for the conveyance of heavy loads up steep elevations: also in railways, &c.

CONVERSATION XX.

Of the Wedge.

FATHER. The next mechanical power is the *wedge*, which is made up of the two inclined planes D E F G and C E F G (Plate IV, Fig. 29), joined together at their bases *e* E F G: D C is the whole thickness of the wedge at its back A B C D, where the power is applied, and D F and C F are the length of its sides; now there will be an equilibrium between the power impelling the wedge downward, and the resistance of the wood, or other substance acting against its sides, when the thickness D C of the wedge is to the length of the two sides, or, which is the same thing, when half the thickness D E of the wedge at its back is to the length of D F one of its sides, as the power is to the resistanee.

Charles. This is the principle of the inclined plane.

Father. It is, and notwithstanding all the disputes which the methods of calculating the advantage gained by the wedge have occasioned, I see no reason to depart from the opinion of those who consider the wedge as a double inclined plane.

Emma. I have seen people cleaving wood with wedges, but they seem to have no effect, unless great force and great velocity are also used.

Father. No, the power of the

attraction of cohesion, by which the parts of wood stick together, is so great, as to require a considerable *momentum* to separate them. Did you observe nothing else in the operation worthy of your attention?

Charles. Yes, I also took notice that the wood generally split a little below the place to which the wedge reached.

Father. This happens in cleaving most kinds of wood, and then the advantage gained by this mechanical power, must be in proportion as the length of the sides of the cleft in the wood is greater than the length of the whole back of the wedge. There are other varieties in the action of the wedge; but, at present, it is not necessary to refer to them.

Emma. Since you said that all instruments, which sloped off to an edge on one side only, were to be explained by the principle of the inclined plane; so, I suppose, that those which decline to an edge on both sides, must be referred to the principle of the wedge.

Father. They must, which is the case with many chisels, and almost all sorts of axes, nails, bayonets, &c.: the teeth of animals act also as wedges. A saw is a series of wedges, on which the motion impressed is oblique to the resistance.

Charles. Is the wedge much used as a mechanical power?

Father. It is of great importance in a vast variety of cases, in which the other mechanical powers are of no avail; and this arises from the momentum of the blow, which is greater, beyond comparison, than the application of any dead weight or pressure, such as is employed in the other mechanical powers. Hence it is used in splitting wood, rocks, &c., and even the largest ship may be raised to a small height by driving a wedge below it.

Emma. Has it been applied to any other purposes?

Father. It is used for raising the beams of a house, when the floor gives way, by reason of too great a burden having been laid upon them.

It is usual also in separating large mill-stones from the siliceous sand-rocks, in some parts of Derbyshire, to bore horizontal holes under them in a circle, and fill these with pegs or wedges made of dry wood, which gradually swell by the moisture of the earth, and in a day or two lift up the mill-stone without breaking it; to this practice Dr. Darwin alludes:—

Climb the rude steeps, the granite cliffs surround,
Pierce with steel points, with wooden *wedges* wound.

The principle of the wedge is called into action by almost every mechanic, and in a thousand instances, in which the reason of the thing is not even thought of. To mention a single instance, builders, in raising their scaffolds, always tighten the ropes round their scaffolding poles by means of wedges driven between the cords and the poles.

CONVERSATION XXI.

Of the Screw.

FATHER. Let us now examine the properties of the sixth and last mechanical power, the *screw*; which, however, cannot be called a simple mechanical power, since it is never used without the assistance of a lever or winch; by which it becomes a compound engine, and it is of great power in pressing bodies together, or in raising great weights. A B (Plate IV, Fig. 30) is the representation of one, together with the lever D F.

Emma. You said just now, papa,

that all the mechanical powers were reducible either to the lever or inclined plane: how can the screw be referred to either?

Father. The screw is composed of two parts, one of which, A B, is called the screw, and consists of a spiral protuberance, called the *thread*, which may be supposed to be wrapt round a cylinder; the other part, C D, called the *nut*, is perforated to the dimensions of the cylinder; and in the internal cavity is also a spiral groove adapted to receive the thread. Now, if you cut a slip of writing-paper in the form of an inclined plane *a b c* (Fig. 30), and then wrap it round a cylinder of wood, you will find that it makes a spiral answering to the spiral part of the screw; moreover, if you consider the ascent of the screw, it will be evident that it is precisely the ascent of an inclined plane.

Charles. By what means do you calculate the advantage gained by the screw?

Father. There are, at first sight, evidently two things to be taken into consideration; the first is the distance between the threads of the screw;—and the second is the length of the lever.

Charles. Now I comprehend pretty clearly how it is an inclined plane, and that its ascent is more or less easy as the threads of the spiral are nearer or farther distant from each other.

Father. Well then let me examine, by a question, whether your conceptions be accurate: suppose two screws, the circumferences of whose cylinders are equal to one another; but in one, the distance of the threads to be an inch apart; and that of the threads of the other only one-third of an inch; what will be the difference of the advantage gained by one of the screws over the other.

Charles. The one, whose threads are three times nearer than those of the other, must, I should think, give three times the most advantage.

Father. Give me the reason for what you assert.

Charles. Because, from the principle of the inclined plane, I learnt that if the *height* of two planes were the same, but the length of one, twice, thrice, or four times greater than that of the other, the mechanical advantage gained by the longer plane would be two, three, or four times more than that gained by the shorter. Now, in the present case, the height gained in both *screws* is the same, one inch, but the space passed in that, three of whose threads go to an inch, must be three times as great as the space passed in the other; therefore, as space is passed, or time lost, just in proportion to the advantage gained, I infer that three times more advantage is gained by the screw, the threads of which are one third of an inch apart, than by that whose threads are an inch apart.

Father. Your inference is just, and naturally follows from an accurate knowledge of the principle of the inclined plane. But we have said nothing about the lever.

Charles. This seems hardly necessary, it being so obvious to any one, who will think a moment, that power is gained by that, as in levers of the first kind, according to the length F D from the nut.

Father. Let us now calculate the advantage gained by a screw, the threads of which are half an inch distance from one another, and the lever 7 feet long.

Charles. I think you once told me that, if the radius of a circle was given, in order to find the circumference, I must multiply that radius by 6.

Father I did; for though that is not quite enough, yet it will answer all common purposes, till you are a little more expert in the use of decimals.

Charles. Well, then, the circumference of the circle made by the revolution of the lever will be 7 feet, multiplied by 6, which is 42 feet or 504 inches; but, during this revolution, the screw is raised only half an inch, therefore the space passed by the moving power will be 1008 times greater than that gone through by the weight, consequently the advantage gained is 1008, or one pound applied to the lever will balance 1008 pounds acting against the screw.

Father. You perceive that it follows as a corollary from what you have been saying, that there are two methods by which you may increase the mechanical advantage of the screw.

Charles. I do; it may be done either by taking a long lever, or by diminishing the distance of the threads of the screw.

Father. Tell me the result then, supposing the threads of the screw so fine as to stand at the distance of but one quarter of an inch asunder; and that the length of the lever were 8 feet instead of 7.

Charles. The circumference of the circle made by the lever will be 8 multiplied by 6, equal to 48 feet or 576 inches, or 2304 quarter inches, and as the elevation of the screw is but one quarter of an inch, the space passed by the power will, therefore, be 2304 times greater than that passed by the weight, which is the advantage gained in this instance.

Father. A child, then, capable of moving the lever sufficiently to overcome the friction, with the addition of a power equal to one pound, will be able to raise 2304 pounds, or something more than 20 hundred weight and a half. The strength of a powerful man would be able to do 20 or 30 times as much more.

Charles. But I have seen, at Mr. Wilmot's paper-mills, to which I once went, six or eight men use all their strength in turning a screw, in order to press out the water of the newly made paper. The power applied in that case must have been very great indeed.

Father. It was; but I dare say that you are aware that it cannot be estimated by multiplying the power of one man by the number of men employed.

Charles. That is, because the men standing by the side of one another, the lever is shorter to every man the nearer he stands to the screw, consequently though he may exert the same strength, yet it is not so effectual in moving the machine, as the exertion of him who stands nearer to the extremity of the lever.

Father. The true method, therefore, of calculating the power of this machine, aided by the strength of these men, would be to estimate accurately the power of each man according to his position, and then adding all these separate advantages together for the total power gained.

Emma. A machine of this kind, is, I believe, used by bookbinders, to press the leaves of the books together before they are stitched?

Father. Yes, it is found in every bookbinder's workshop, and is particularly useful where persons are desirous of having small books reduced to a still smaller size for the pocket. It is also the principal machine used for coining money, for taking off copper-plate prints, and for printing in general.

Charles. I remember Dr. Darwin's description of coining.

With iron lips his rapid rollers seize
The length'ning bars, in thin expansion
    squeeze;
Descending *screws* with pondrous fly-wheels
    wound
The tawny plates, the new medallions round;
Hard dies of steel the cupreous circles cramp,
And with quick fall his massy hammers stamp,
The Harp, the Lily, and the Lion join,
And GEORGE and BRITAIN guard the sterling
    coin. BOTANIC GARDEN.

Father. These lines are descriptive of Mr. Boulton's magnificent apparatus for coining; the whole machinery is worked by an improved steam-engine, which rolls the copper for halfpence, works the screw-presses for cutting out the circular pieces of copper, and coins both the faces and edges of the money at the same time: and, since the circulation of the new halfpence, we are all acquainted with the superior excellence of the workmanship. By this machinery, four boys ten or twelve years old are capable of striking 30,000 guineas in an hour, and the machine itself keeps an unerring account of the number of pieces struck. A higher treat an inquisitive youth cannot have, than that of witnessing the process of coining as it is carried on at the Mint, Tower Hill.

Emma. And I have seen the

cyder-press in Kent, which consists of the same kind of machine.

Father. It would, my dear, be an almost endless task, were we to attempt to enumerate all the purposes to which the screw is applied in the mechanical arts of life: it will, perhaps, be suffiicient to tell you, that wherever great pressure is required, there the power of the screw is uniformly employed.

END OF THE FIRST VOLUME.

TO THE BINDER.

Please to place the Plates at the end of the Volumes.

CHARLES WOOD, Printer,
Poppin's Court, Fleet Street, London.

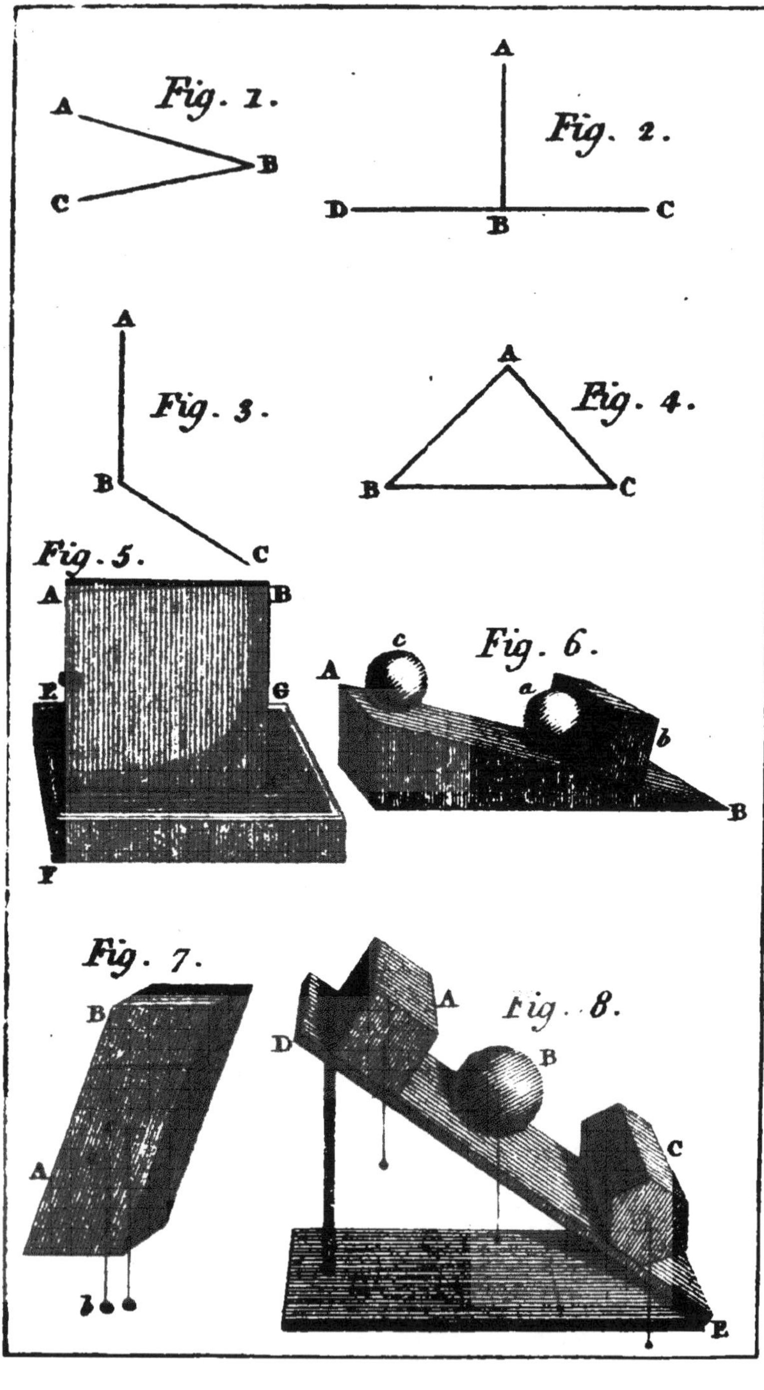
Fig. 1.
A
B
C
Fig. 2.
A
D
B
C
Fig. 3.
A
B
C
Fig. 4.
A
B
C
Fig. 5.
A
B
E
G
F
Fig. 6.
c
A
a
b
B
Fig. 7.
B
A
b
Fig. 8.
A
D
B
C
E

Fig. 9.

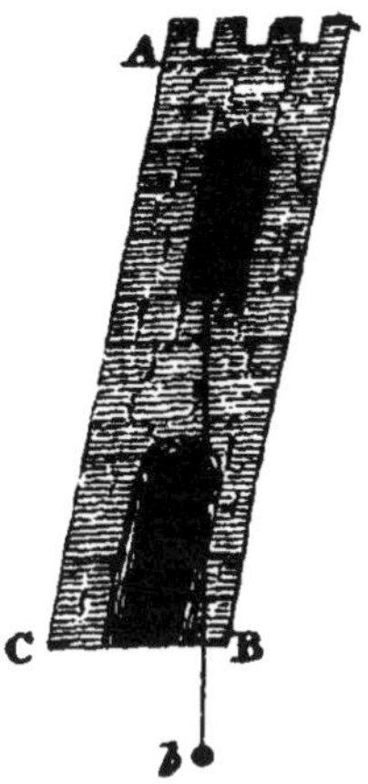

Fig. 10.

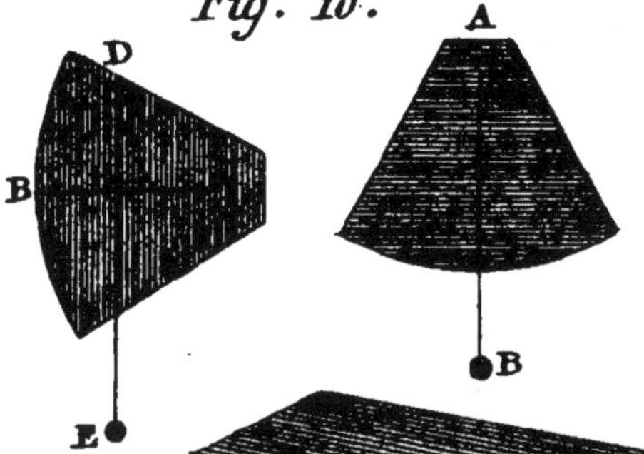

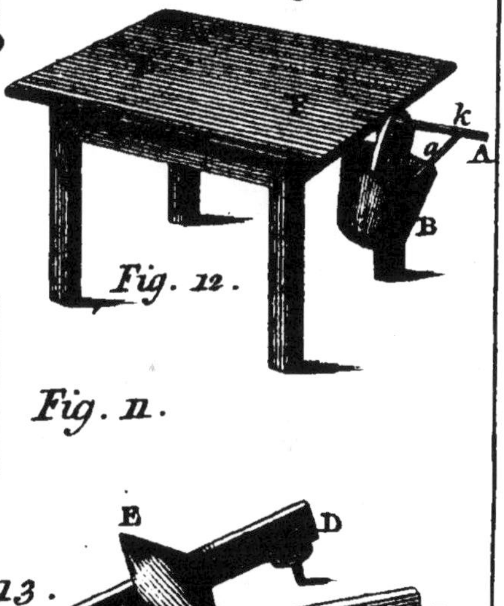

Fig. 12.

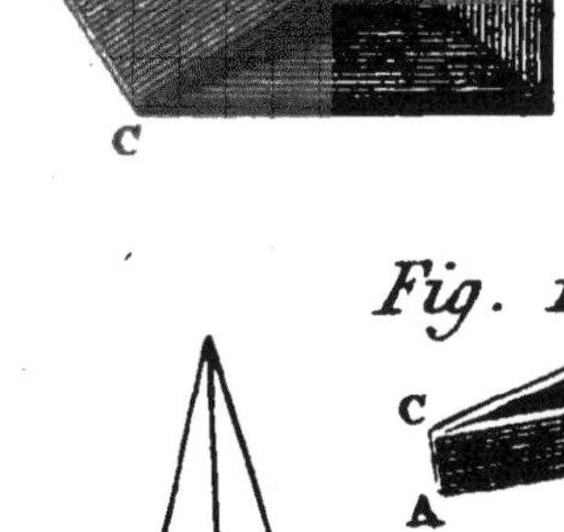

Fig. 11.

Fig. 13.

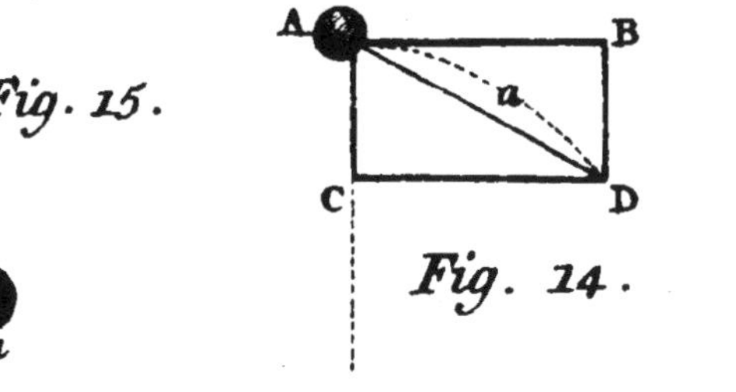

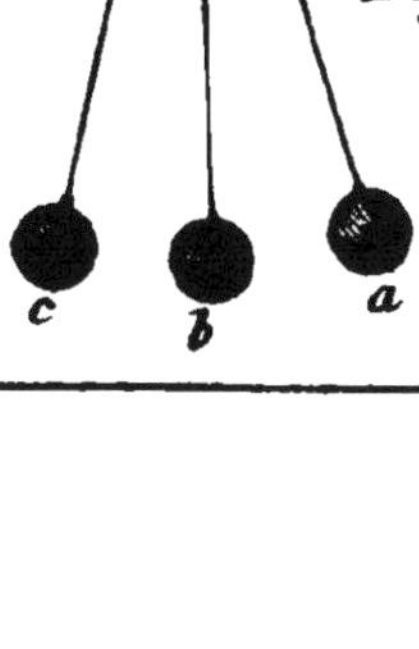

Fig. 15.

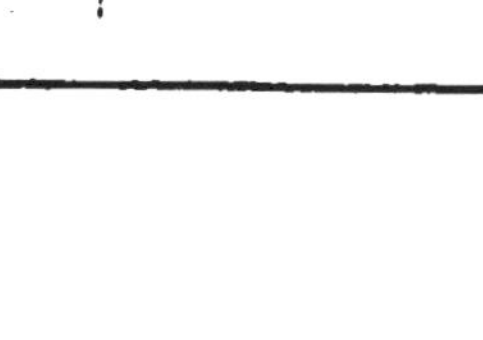

Fig. 14.

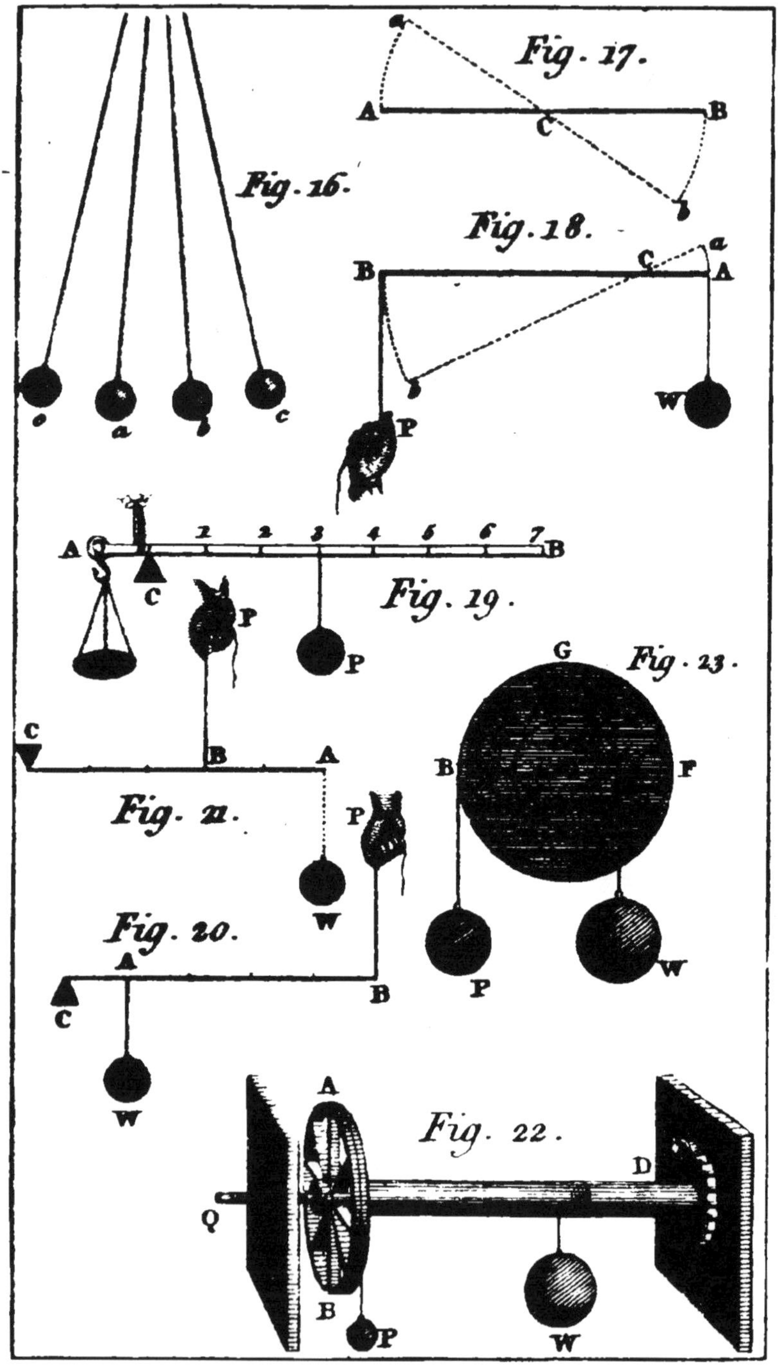
Fig. 16.
o
a
b
c
Fig. 17.
a
A
C
B
b
Fig. 18.
B
C
a
A
b
P
W
A
1
2
3
4
5
6
7
B
C
Fig. 19.
P
P
G
Fig. 23.
B
F
P
W
C
B
A
Fig. 21.
P
W
Fig. 20.
A
C
B
W
A
Fig. 22.
D
Q
B
P
W

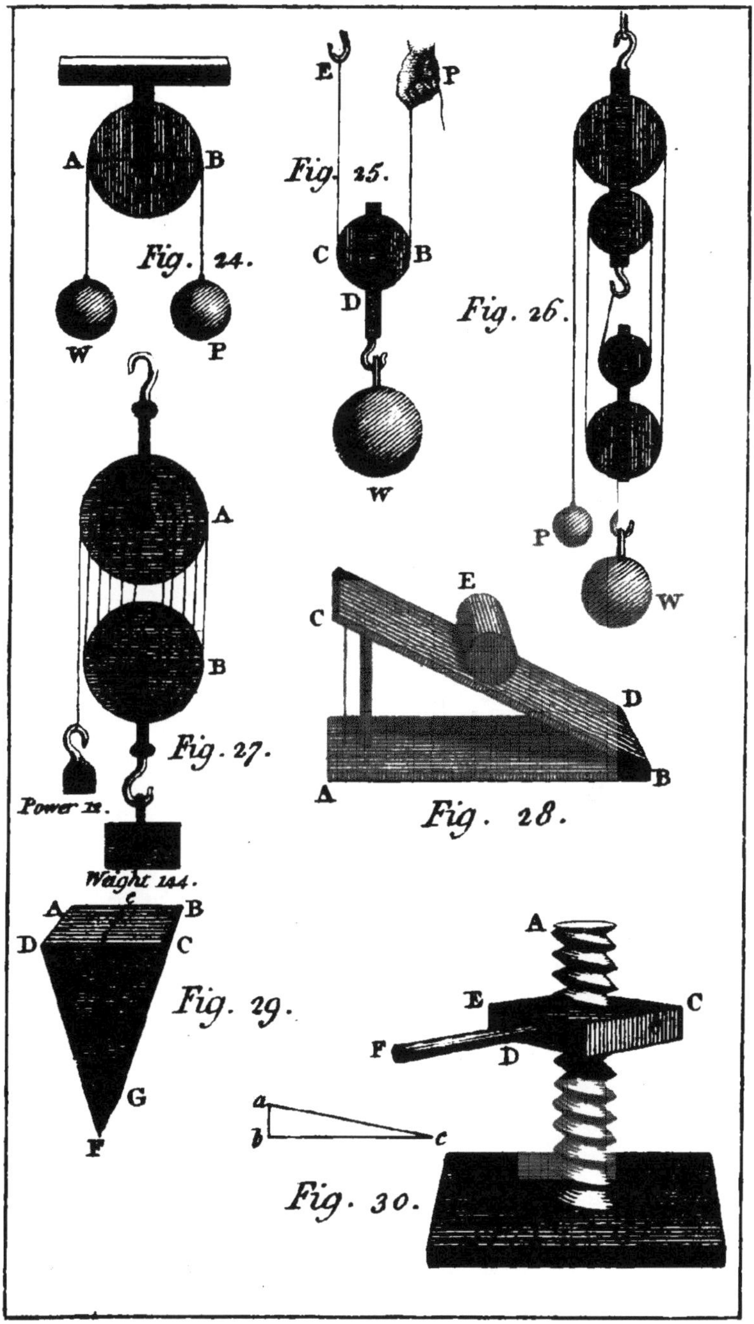
A
B
Fig. 24.
W
P
E
P
Fig. 25.
C
B
D
W
Fig. 26.
P
W
A
B
Fig. 27.
Power 12.
Weight 144.
E
C
D
A
B
Fig. 28.
A
c
B
D
C
Fig. 29.
G
F
A
E
C
F
D
a
b
c
Fig. 30.

Spotted something that isn't quite right?

Let us know and we'll fix it for future reprints. And as a thank you for your diligence, we'll send you a free Sothis book of your choice.

publisher@sothis.press

This page intentionally left blank

Made in the USA
Coppell, TX
05 November 2026

11327621R00129